The Energy Transition

Edward Cussler · Abhoyjit S. Bhown

The Energy Transition

A Primer

Edward Cussler
(Emeritus), Department
of Chemical Engineering
University of Minnesota
Minneapolis, MN, USA

Abhoyjit S. Bhown
Electric Power Research Institute
Palo Alto, CA, USA

ISBN 978-3-032-06640-4   ISBN 978-3-032-06641-1   (Ebook)
https://doi.org/10.1007/978-3-032-06641-1

© The Editor(s) (if applicable) and The Author(s), under exclusive license to Springer Nature Switzerland AG 2026

This work is subject to copyright. All rights are solely and exclusively licensed by the Publisher, whether the whole or part of the material is concerned, specifically the rights of translation, reprinting, reuse of illustrations, recitation, broadcasting, reproduction on microfilms or in any other physical way, and transmission or information storage and retrieval, electronic adaptation, computer software, or by similar or dissimilar methodology now known or hereafter developed.
The use of general descriptive names, registered names, trademarks, service marks, etc. in this publication does not imply, even in the absence of a specific statement, that such names are exempt from the relevant protective laws and regulations and therefore free for general use.
The publisher, the authors and the editors are safe to assume that the advice and information in this book are believed to be true and accurate at the date of publication. Neither the publisher nor the authors or the editors give a warranty, expressed or implied, with respect to the material contained herein or for any errors or omissions that may have been made. The publisher remains neutral with regard to jurisdictional claims in published maps and institutional affiliations.

The views and conclusions expressed in this book are solely those of the authors and do not necessarily reflect the views and conclusions of their employers or affiliated institutions.

This Springer imprint is published by the registered company Springer Nature Switzerland AG
The registered company address is: Gewerbestrasse 11, 6330 Cham, Switzerland

If disposing of this product, please recycle the paper.

Preface

Our society now accepts the need to make major changes, especially in energy use. Ironically, the acceptance of this need results from the COVID pandemic. In 2019, before COVID, many would ask us if there was going to be an energy crisis. Now, most ask us how bad the energy crisis will be.

Our new society cannot use fossil fuels like coal, oil, and gas so profligately without controlling their emissions. It will depend more on energy-saving devices like heat pumps and LED bulbs for home heating and lighting. In our new society, traditional cars may no longer be as central, and gas stations will be rarer. There may be a carbon tax: like the excise tax on booze, it may be broadly accepted. After all, in the early 18th century, there was such an epidemic of alcoholism—Hogarth's "gin craze"—that one could be drunk for a penny. This threatened society so much that the entire public accepted new anti-drinking laws. Today, we cannot get drunk for a penny, but we can drive for pennies.

We will need a new social design reflecting the total cost of driving, including gasoline and environmental effects.

This major social redesign is too important to be left to energy experts like us and our peers. Instead, we need to return to the ideals of Thomas Jefferson, who argued that the key to a successful society was a common cultural literacy, with a population that shared traditions, regular activities, and political knowledge of our common culture. This might include ideas about voting, why New Year's Day is significant, which "truths are self-evident", and who Little Red Riding Hood is. Such ideas are basic to a common cultural literacy.

We need to develop a scientific literacy that parallels this cultural literacy. This scientific literacy should be known by the public. However, the public seems scientifically unaware. This unawareness is partly the fault of those technically trained educators who, like us, have pretended that we have scientifically educated our students with watered-down, rarely respected science classes. The students often agree that these classes have limited value and use derogatory titles to describe them. They have felt justified because the content of these courses is perceived as dilute.

In this book, we describe a scientific literacy for energy. Energy, the core of the industrial revolution centuries ago, will continue to drive our new society. Fossil fuels—coal, oil, and natural gas—which are convenient, ubiquitous, and store energy, have raised the standard of living for all society and lifted billions out of poverty. But they are environmentally detrimental, so we must change how our industrialized society operates.

This book begins to develop this scientific literacy. However, the book is not about environmental advocacy. It doesn't assign blame for our current situation or urge specific changes to reduce temperature increases or to delay climate-induced immigration. It discusses what options we have to make our energy-dependent society less dependent on fossil fuels. It describes controlling their emissions when society does use them. Specifically, the book does talk about the following:

- *In what ways do we currently use energy?*
  The book reports how much energy we have already used and the trends in our consumption.
- *Is sustainable energy even possible?*
  The book reviews sustainable energy for electricity and transportation, but it also shows how energy is central to manufacturing plastics, detergents, and drugs. It shows how producing even one pork chop releases a kilogram of carbon dioxide.
- *How is energy use described?*
  The book defines and explores "energy balances." Balancing energy, roughly like balancing a checkbook, is tricky and has mystified many, beginning with coal mine owners in the 18th century.
- *Can we rely on sustainable energy sources like the sun and wind?*
  Will the lights go out and the furnace shut off when the sun goes down and the wind is calm? How will the cost compare to traditional sources?

Finally, the book describes options for what future energy sources could look like.

The book assumes some mathematical knowledge, but not much beyond that used in balancing a checkbook. It assumes some chemical knowledge but not much beyond knowing that CO_2 is carbon dioxide. Again, it does not say what we should do about sustainable energy; rather, it discusses what we could do.

Minneapolis, USA
Palo Alto, USA

Edward Cussler
Abhoyjit S. Bhown

Contents

Part I Where We Are

1 Energy Demand   3
  1.1 What We Need to Supply   8
  1.2 How We Describe Energy   11
  1.3 Why Distinguishing Work from Heat Is Hard (Important and Difficult, but Short)   17
  1.4 How We Estimate the Future   20
  1.5 Conclusions   24
  References   25

2 Unsustainable Energy: Coal   27
  2.1 What Coal Is   28
  2.2 Where Coal Is   31
  2.3 What Coal Is Used for   33
  2.4 Reducing Emissions by Carbon Capture and Storage (CCS)   36
  2.5 Biomass, a Tangent   37
  2.6 Conclusions   41
  References   41

3 Unsustainable Energy: Oil and Gas 43
- 3.1 What Natural Gas Is 45
- 3.2 What Crude Oil Is 48
- 3.3 How Much Oil and Gas Are Left 52
- 3.4 Oil and Gas Processing 56
  - 3.4.1 Two Key Processing Steps 56
  - 3.4.2 What Processed Oil Is Used for 60
  - 3.4.3 Making Hydrogen from Natural Gas 64
- 3.5 Hydrogen and Ammonia as Fuels 65
- 3.6 Conclusions 67
- References 67

Part II Towards Where We Need to Be

4 Energy Transformations (The Hard Part of the Book) 71
- 4.1 Overview of Physical Energy Balances 73
  - 4.1.1 Electrical Energy 74
  - 4.1.2 Potential Energy 75
  - 4.1.3 Kinetic Energy 75
  - 4.1.4 Energy Conversion 76
- 4.2 Genesis of Thermodynamics 78
  - 4.2.1 Sadi Carnot (1796–1832) 79
  - 4.2.2 Rudolf Clausius (1822–1888) 81
  - 4.2.3 Ludwig Boltzmann (1844–1906) 82
- 4.3 Converting Heat to Work 83
  - 4.3.1 The Carnot Cycle 84
  - 4.3.2 Engine Efficiency 89
  - 4.3.3 Heat Pumps 90
- 4.4 Entropy and Energy 91
  - 4.4.1 Entropy as Disorder 92
  - 4.4.2 Gibbs Free Energy 93
  - 4.4.3 Energy of Unmixing 95

	4.5	Chemical Energy Balances	98
		4.5.1 Combustion	98
		4.5.2 Electrochemical Energy	101
	4.6	Conclusions	102
		Reference	103
5	**Is Sustainable Energy Feasible? Steady Energy Resources**		105
	5.1	Why Energy Conservation Is Hard to Understand	106
	5.2	Sustainable Energy Is Cheap	109
	5.3	Hydropower Can Be Steady	114
	5.4	Nuclear Power Can Be (Nearly) Steady	116
	5.5	Conclusions	119
		References	120
6	**Is Sustainable Energy Feasible? Periodic Energy Resources**		121
	6.1	Solar Is Ready Now	122
		6.1.1 Silicon Solar Cells	122
		6.1.2 Solar Heating for Domestic and Commercial Use	126
		6.1.3 Solar Energy for Chemicals	127
	6.2	Wind Is Ready Now	128
	6.3	Conclusions	130
		Reference	131
7	**Sustainable Energy Requires Energy Storage**		133
	7.1	Mechanical Energy Storage	134
		7.1.1 Storing Energy with Water	134
		7.1.2 Storing Energy as Compressed Gas	137
	7.2	Thermal Energy Storage	138
		7.2.1 Sensible Heat Storage	139
		7.2.2 Latent Heat Storage	142

	7.3	Chemical Energy Storage	143
		7.3.1 Hydrogen	144
		7.3.2 Ammonia	147
		7.3.3 Liquid Organic Hydrogen Carriers (LOHCs)	148
	7.4	Electrochemical Energy Storage	149
		7.4.1 Lithium-Ion Batteries	151
		7.4.2 Lead-Acid Batteries	153
		7.4.3 Vanadium Redox Batteries	154
	7.5	Considerations for Energy Storage	156
	7.6	Conclusions	158
	Reference		159
8	**Carbon Capture and Storage**		**161**
	8.1	California's Duck Curve	162
	8.2	Carbon Capture and Storage (CCS)	164
		8.2.1 Carbon Capture	165
		8.2.2 CO_2 Storage	168
	8.3	Where CCS Has Promise	170
	8.4	Direct Air Capture	173
	8.5	Conclusions	175
	References		175
9	**The Energy Transition**		**177**
	9.1	Is Climate Change Real?	179
	9.2	A Decarbonized Path Forward	181
	9.3	Energy for the Individual	184
		9.3.1 Saving Energy at Home	184
		9.3.2 Saving Energy at Dinner	186
	9.4	How to Test What You've Learned	189
	9.5	Conclusions	192
	References		196

About the Authors

Edward Cussler, Distinguished Institute Professor at the University of Minnesota, was educated at Yale (BE, 1961) and Wisconsin (Ph.D., 1965). After 13 years at Carnegie-Mellon University, Cussler joined the University of Minnesota in 1980. He has written over 250 articles and five books, including *Diffusion*, *Bioseparations*, and more recently, *Chemical Product Design*. Cussler, who served the American Institute of Chemical Engineers (AIChE) as President, was awarded the Colburn (1975) and Lewis (2001) awards, and was the Institute Lecturer (2014). He holds honorary doctorate degrees from the Universities of Lund (2002) and Nancy (2007). Cussler is a Fellow of the American Association for the Advancement of Science and a member of the National Academy of Engineering. His current research interest is the production of ammonia from stranded wind energy, that is, "putting wind energy in a bottle." This book came to fruition thanks to the encouragement of his wife, Betsy, and their children.

Abhoyjit S. Bhown is a Senior Program Manager at the Electric Power Research Institute (EPRI) in Palo Alto, California. He holds degrees in chemical engineering (BS 1982, Auburn; MS 1983, California Davis; and Ph.D. 1990, Minnesota) and in mathematics (MS 1989, Minnesota). At EPRI since 2007, he leads a team of scientists and engineers focused on carbon management, including novel power cycles, CO_2 capture, and CO_2 storage. The effort spans computational, laboratory, bench, field pilot studies, and engineering designs. Prior to EPRI, he founded and led start-up companies for separations technologies and worked at research institutes and energy companies. Most of his research has focused on separations technologies. The support of his wife, Nivedita, and their children, Anirudh and Ashutosh, made this book possible.

List of Figures

Fig. 1.1 U.S. energy consumption in 2024. Power use is about 10 kW per person. This use produces 40 kg CO_2 per person per day. The values given here are in exajoules (EJ) or 10^{18} J. (Based on U.S. Energy Information Administration 2025a)   5

Fig. 1.2 Coal production versus time. Production starts slowly, accelerates, and then slows as the coal mine is exhausted   21

Fig. 1.3 Three approximate estimates of the amount produced. In **a**, the rate is constant till exhaustion; in **b**, it rises exponentially to a limit; and in **c**, it follows a logistic curve   22

Fig. 2.1 **Coal production in the U.S.** The production of coal in the United States is in sharp decline (Data from The Energy Institute 2025)   32

Fig. 2.2	**Coal Consumption by Several Large Countries.** China's major role in coal is complemented by it having the world's largest increase in solar-generated electricity (Data from The Energy Institute 2025)	34
Fig. 2.3	**Major Uses of Coal in the U.S.** Coal is largely used to make electricity. Other uses are steel and cement (U.S. Energy Information Administration 2023b)	34
Fig. 2.4	**How coal is burned to produce electricity.** Coal is burned to make steam, which rotates the blades of a turbine. The turbine turns a generator to make electricity (TVA 2025)	35
Fig. 3.1	**Natural Gas Combined Cycle Power Plant.** Gas turbines and steam turbines are used to make electric power from natural gas	47
Fig. 3.2	**Producing Crude Oil with Steam Injection.** The steam reduces oil viscosity, facilitating flow (U.S. Department of Energy 2008)	50
Fig. 3.3	**Where oil and gas are located.** Most proven reserves of oil and gas are in the Middle East (Xu and Bell-Hammer 2023)	52
Fig. 3.4	**How fracking works.** A slurry of soapy water and sand is pumped into an existing well to fracture the rock, releasing gas and oil (European Environment Agency 2021)	55
Fig. 3.5	**Products made from crude oil.** These are dominated by fuels, but the greatest value can be in chemicals (Data from U.S. Energy Information Administration 2024)	57

List of Figures

Fig. 4.1	**The Carnot cycle.** This frictionless cyclic engine, of two isothermal steps and two adiabatic steps, has an efficiency much less than one hundred percent	84
Fig. 4.2	**Two equivalent ways to show a Carnot cycle.** The pressure P—volume V plot on the left is due to Carnot; the temperature T—entropy S plot on the right is due to Clausius	86
Fig. 4.3	**A heat engine.** A heat engine converts thermal energy into work	89
Fig. 5.1	**Energy use, including some issues of heat energy and work energy.** U.S. Energy consumption totaled 99.4 EJ in 2024, or about 10 kW per person (Data from U.S. Energy Information Administration 2024)	108
Fig. 5.2	**Energy costs versus installed capacity from 2010 to 2019.** As coal capacity increases, its cost per megawatt does not drop. As solar capacity increases, its cost drops dramatically (Roser 2025)	113
Fig. 6.1	**Electron energy versus photon (light) energy.** After a threshold is reached, current can flow	124
Fig. 6.2	**A Schematic picture of a silicon-based solar cell. a** Light which reaches the interface across which a potential is applied produces a current. **b** Thus, light generates electricity (ACS ChemMatters 2014)	125
Fig. 7.1	**Some Materials Used to Store Hydrogen.** The dashed region is the target for commercialization (DOE 2024)	146
Fig. 7.2	**The basic structure of a lithium-ion battery.** The energy comes from lithium-ion energy reduction	152

Fig. 7.3	**A schematic diagram of a lead-acid battery drawing current**. During discharge, the lead is converted to a lower energy. During charging, the flow of electrons is reversed	154
Fig. 7.4	**A vanadium redox battery**. It has a lower energy density but charges and discharges faster because it does not involve diffusion in solids	155
Fig. 8.1	**California's Duck Curve**. The amount of power generated by non-renewable sources drops to near zero during daylight hours and curves steeply at dawn and dusk (U.S. Energy Information Administration 2023)	163
Fig. 8.2	**Carbon capture**. A solvent circulates between two columns, capturing CO_2 in the absorber column and releasing it in the stripper column	167
Fig. 8.3	**Minimum thermodynamic work for capturing CO_2 at 40 °C and compressing it to 150 bar**. Separations of more dilute mixtures of CO_2 cost more energy per mass of CO_2	171
Fig. 8.4	**CO_2 storage**. **a** CO_2 is compressed to a high-density supercritical fluid and **b** stored underground in suitable geological formations (based on CO2CRC 2017; US Congressional Budget Office 2023)	171
Fig. 9.1	**Carbon dioxide concentrations for the last 800,000 years (NOAA 2024a)**. The recent increase in CO_2 concentration is a result from human activity	180

Fig. 9.2	**Energy Generation Pathway to 1.5 °C and Net Zero by 2050 (IEA 2023).** Integration Assessment Models can provide scenarios that result in **a** lowest cost energy generation to **b** achieve net zero emissions	182
Fig. 9.3	**Domestic Energy Use in the United States.** About 75% of the energy used in our homes is for heating and cooling (DOE 2024)	185
Fig. 9.4	**Emissions from gasoline-powered and electric cars.** The environmental impact is more complicated than first expected (Carbon Brief 2020)	191
Fig. 9.5	**Global and US energy consumption.** **a** The rise in global energy consumption has outpaced the rise in global renewable energy. **b** The US energy consumption has been largely flat for decades, with increasing amounts of renewables. Overall, both global and US energy is over 80% fossil fuel (Data from Energy Institute 2024)	194

List of Tables

Table 1.1	Average power consumption per person in different countries	8
Table 1.2	Carbon dioxide emissions from different fossil fuels	10
Table 1.3	Common energy units	14
Table 1.4	Common conversion for energy units	15
Table 1.5	Common units for power	17
Table 2.1	Heats of combustion of biomass	39
Table 3.1	Heats of combustion of some fuels	61
Table 3.2	Chemicals from oil and gas in the U.S.	63
Table 4.1	Heats of combustion for some fuels	99
Table 4.2	Energy densities of common batteries	102
Table 5.1	Capital and operating costs for different ways to make electricity	111
Table 7.1	Hydroelectric energy storage	136
Table 7.2	Heat capacity of materials	139
Table 7.3	Comparing common and less common fuels	144
Table 7.4	Batteries for energy storage	150
Table 7.5	Round-trip efficiency	157

Table 9.1	Infrastructure needed to make 50 EJ of low-carbon energy by 2050	184
Table 9.2	The carbon footprint of food	188

Part I

Part I Where We Are

1

Energy Demand

Our society ultimately uses energy in two ways—heat and work. We use heat to keep ourselves warm in our homes, cook our food, and manufacture things like steel. But sometimes we want to move ourselves or displace things from one location to another, which also requires energy. We refer to this use of energy for movement as work. It takes work, not heat, to move things like bushels of corn, water from a well, or even us. Though our ancestors harnessed heat ever since they controlled fire over a million years ago, it was the Industrial Revolution in the eighteenth century that truly ushered in widespread reliance on work. Existing machines were improved, new ones were invented, and energy was converted from heat to work. The most significant machine was the steam engine, where coal or oil was burned to generate heat, boil water, move a piston attached to a crankshaft, and rotate a wheel. Such heat-to-work conversion machines, called engines, significantly reduced the burden of human and animal labor, ushering in global prosperity. They remain the backbone of our modern civilization.

Energy is measured in the same units regardless of whether it is heat or work, joules (J) being a common unit. For example, taking a 10-min hot shower using a low-flow shower uses around 10 megajoules (MJ or 10^6 J) of energy. This is the quantity of heat needed to warm up about 80 L of water by about 30 °C or about 50 °F.

Work also has units of joules but implies movement. If you pick up a medium-sized 100 g tomato from the ground to one meter height, you'll spend about one joule of energy. Do the same tomato-picking activity on the moon, where gravity is smaller, and you'll use just 0.16 J of energy.

Power, which is defined as energy divided by time, is a measure of how fast energy is consumed or generated. One watt (W) of power is equal to one joule of energy divided by one second. If the tomato-lifting activity is completed in one second, you have used a power of 1 W. If the same activity takes 10 s, you have used a power of 0.1 W. The energy you've used is still 1 J or equivalently 1 W-s (watt-second). Confusingly, though heat and work have the same units, they represent different concepts: heat is energy transferred because of temperature differences, while work is energy transferred through forces and displacement.

The world's energy is used as heat directly or as heat converted into work using machines like gasoline-powered cars. Most of this energy is made from non-renewable resources—particularly fossil fuels like oil, gas, and coal, as shown in Fig. 1.1 for the U.S. (U.S. Energy Information Administration 2025a). To examine this situation in more detail, consider the current energy source shown on the left-hand side of the figure. In the U.S., we are consuming 99.4 exajoules (EJ or 10^{18} J) of energy every year. Of this energy, 37% currently comes from oil, and 36% from natural gas. While coal was a major energy source in the past, it is now

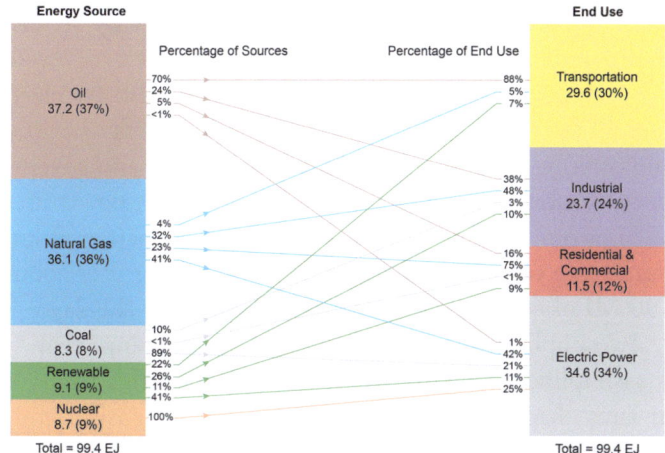

Fig. 1.1 U.S. energy consumption in 2024. Power use is about 10 kW per person. This use produces 40 kg CO_2 per person per day. The values given here are in exajoules (EJ) or 10^{18} J. (Based on U.S. Energy Information Administration 2025a)

only about 8% of our total use. At present, only around 9% of our energy comes from renewable sources, and the remaining 9% comes from nuclear power. The U.S. energy consumption has been largely flat for the past 20 years, though the world's energy consumption has grown by 25% during that time. We note, however, that the projected growth of artificial intelligence and data centers is expected to significantly increase energy demand in the United States and other parts of the world.

On a per capita basis, annual energy consumption is about 300 GJ (10^9 J) in the U.S., and about 75 GJ in the world (Statistical Review of World Energy 2025). Averaged over the course of a year, this translates to a per capita power consumption of about 10,000 J/s or 10 kW for the U.S. and about 2500 J/s or 2.5 kW for the world.

The ways in which we use this energy are shown on the right-hand side of this figure. Transportation accounts for 30%, not only by personal cars but also by trucks, trains, airplanes, and ships. Industry uses 24% to make not only heavy chemicals like concrete but also fertilizer, central to food production. Residential use, like heating and cooking, and commercial use, like business and government combined, is about 12%. Making electric power uses 34% of our energy. To make this more personal, a 60 km round-trip commute in a gasoline-powered car uses about 190 MJ of energy—about 2 kW averaged over the day.

Two implications of Fig. 1.1 are worth stressing. First, it is a simplified diagram; we shall build upon this diagram in later chapters to provide a more comprehensive landscape of energy. Second, as shown on the left-hand side of the diagram, most of the energy generated is thermal. Oil, natural gas, and coal are burned in air to generate carbon dioxide, water, and heat. Nuclear reactions also release heat that turns water into steam. Renewables like solar and wind do not release heat but instead make electricity directly for doing work. On the right side of the diagram, transportation, industrial, residential, and commercial sectors all use energy. Some of this energy is used as heat directly, like home heating, while some of the energy is converted from heat to work in machines, like a car. We will talk more about this conversion later in this chapter.

We can also talk about energy in terms of carbon dioxide. We humans produce over 37 billion tonnes (Gt or gigatonnes) of CO_2 per year from energy-related activity (International Energy Agency 2025), of which the U.S. produces about 5 Gt CO_2 per year (U.S. Energy Information Administration 2025b). This is 40 kg of CO_2 per person per day, or 1.7 kg per hour, and includes all human activity. Over the past two decades, U.S. energy consumption has remained flat. At the same time, CO_2 emitted has dropped by nearly 20%, mostly by displacing coal with natural gas in power plants and increasing renewable energy, subjects we shall explore further in later chapters.

Historically, global energy consumption has grown 2–3% each year, over 80% of which has been provided by fossil fuels. Most believe that energy demand will continue to grow as countries and economies develop, demanding more energy. To meet this future demand, we must develop more sources of sustainable energy, that is, energy obtained from the sun, as sunlight or wind. We must consider nuclear energy, now responsible for 9% of our electric power, as an effective means to reduce our reliance on fossil fuels. We will continue to explore many forms of renewable energy, like alcohol made by biomass fermentation. And we must also consider that fossil fuels may still be around and discuss ways to mitigate their emissions with carbon capture and storage.

In this chapter, we will discuss energy demand in more detail. In Sect. 1.1, we compare U.S. use with that in other countries. We discuss variations of energy demand during the day and with changing seasons, and opportunities for energy conservation, ranging from home insulation to efficient light bulbs. In Sect. 1.2, we list the tools we need to describe energy, including units for both energy and power. In Sect. 1.3, we first roughly sketch the complicated conversions between work and heat, an important topic detailed in greater depth in Chap. 4. In Sect. 1.4, we review models that estimate reserves of fossil fuels and how these can influence policy. These basics put us in a position to discuss what resources we have now and how alternatives can be developed.

1.1 What We Need to Supply

In the United States, we use a power of 10 kW per person, averaged over a year. In other words, for every person, we are consuming the energy of 100 100-W light bulbs lit both day and night. This domestic energy demand includes all uses of energy. As shown in Table 1.1, our energy use is higher than that of most other countries (Estimated from The Energy Institute 2025).

In the U.S., about 12% is commercial and residential—for domestic cooking, heating, and cooling. About 30% is for transportation; private cars are the largest fraction, bigger than moving products to and from markets. Our energy use varies with the time of day: it is about 70% larger in the evening than in the middle of the day. Our energy use varies with the time of year: it is about 50% larger in the winter than in the spring. Any new energy grid that we design must include these diurnal and seasonal variations.

Reflection: The Power of 10 kW per person seems to be a lot. See how you think you yourself could use this much.

Table 1.1 Average power consumption per person in different countries

Country	Power per person, kW
U.S	10
UK	3
Spain	3.5
China	3.5
India	0.9
World	2.5

These values include all uses of energy (Estimated from Review 2025)

In particular, how much energy do you use commuting 20 miles to work in a personal car?

Reflection: Suggest ways in which you yourself could reduce your energy footprint. Specific ideas could include heating water with the sun, using cold water for laundry and personal hygiene, and efficiently charging your computer and your phone.

The large amount of energy we currently use may offer opportunities for conservation and for innovation. We can increase the insulation of our homes, which would reduce our heating bills by about 15%. We can replace incandescent light bulbs with the more advanced light-emitting diodes (LEDs). Electric company executives sometimes describe incandescent bulbs as heaters that give off a small amount of light. The LEDs will save 75% of the energy used in lighting. The U.S. Department of Energy expects these changes to result in savings worth less than $225 per year per household (U.S. Department of Energy 2025)—significant but not compelling for most households.

In addition to increasing sustainable energy consumption while meeting society's demand for energy, many scientists, activists, and individuals also emphasize strategies for CO_2 reduction. One such strategy is the adoption of electric cars. They do not directly emit CO_2, but if the electricity for cars is made from fossil fuels, indirect CO_2 emissions still occur where the electric power is generated. The actual emissions depend on the fuel, given below in Table 1.2 as kilograms of CO_2 per billion joules of energy (10^9 J or GJ). Natural gas emits less CO_2 than oil or coal, but the amount emitted is still large.

Another way to control emissions from fossil fuel use is carbon capture and storage (CCS), a topic further explored in Chap. 8. This strategy separates CO_2 from combustion exhaust gas and injects it underground. While this

Table 1.2 Carbon dioxide emissions from different fossil fuels

Fossil fuel	kg $CO_2/10^9$ J
Coal	90
Oil	68
Natural gas	50

is difficult on a small internal combustion engine, it is straightforward on a large fossil-fired power plant that makes electricity. This strategy lets us continue to use fossil fuels but with negligible CO_2 emissions. It increases the cost of electricity by at least 50% and sometimes much more, and it has a manageable risk of storing CO_2 in, for example, underground formations. But as long as society's demand for energy remains high, CCS could give us more time to transition to a sustainable future at the lowest possible cost.

Many have suggested seeking chemical reactions that convert CO_2 and water back into fuels and oxygen:

$$CO_2 + 4H_2 \rightarrow CH_4 + 2H_2O$$

Such reactions, which require energy to move forward, have been extensively studied over the past century. One major oil company did successfully feed a full-sized jet engine with hot CO_2 and steam to make methane. However, running this jet engine backward took an enormous amount of energy, and it is not at all clear where that energy would come from, except from burning more fossil fuels. Of course, the point of the experiment was to avoid burning fossil fuels in the first place.

Despite cautionary experiences like this, a major university recently proposed a new research institute whose principal goal was to search for this magic reaction. This may be intellectually fascinating, but there is limited use in studying the reaction above without identifying a source of sustainable energy at the gigatonne scale. The reaction chemistry isn't the only problem—the energy needed to drive the reaction also is.

1.2 How We Describe Energy

Next, we describe in more detail the analysis of supplying energy from fossil fuels. We will cast this description in terms of two sets of ideas: energy balances and energy units.

Energy balances. "Balances" simply means the amount we have equals the amount we put in minus the amount we take out. Think of a bank balance: the amount of money that we have equals the amount in minus the amount out. The same arithmetic applies to mass: the amount of carbon retained in a human body equals the carbon in minus carbon out. Note this is not true for specific chemical compounds: the amount of glucose in the human body is *not* equal to glucose in minus glucose out because sugar is metabolized in the body.

Energy balances are similar but more complicated because energy has so many different forms. The potential energy of water within the tank at the top of a water tower is bigger than that of the water at the base of the tower. The kinetic energy of a baseball in flight is higher than that of a baseball held in a catcher's mitt. The energy in the stretched rubber band is higher than the energy in an unstretched rubber band.

The energy of specific chemicals is still more complicated. The energy of an atom of carbon in CO_2 is different than the energy of a carbon atom in pure graphitic carbon. These different energies are partially summarized by chemical formulae. However, the chemical formula of glucose $C_6H_{12}O_6$ (from honey) is the same as for galactose $C_6H_{12}O_6$ (from milk); but the energies are different because the sugars have different arrangements of atoms. While we will not discuss these arrangements in this book, we must recognize that the atomic arrangements will affect the energy of the compounds involved. All this will make energy balances more complicated than bank balances.

We are also going to be concerned with chemical reactions. One of the most important reactions in this book, and indeed of all human energy consumption, is the combustion of carbon:

$$C + O_2 \rightarrow CO_2$$

This equation says that an atom of carbon reacts with a molecule of oxygen to make a molecule of CO_2. It also says that a specific number of carbon atoms and oxygen molecules may react as shown. The number of atoms used as a standard is about 6×10^{23}, a quantity called a mole. It seems arbitrary, but it's not. One mole of carbon atoms weighs 12 g; one mole of oxygen atoms weighs 16 g; and one mole of CO_2 molecules weighs 44 g. This picture of chemical reactions is associated with John Dalton (1766–1844), who was the son of a weaver and who was prohibited from going to a university because he was a Quaker.

We could also write this reaction as:

$$12\ C + 32\ O_2 \rightarrow 44\ CO_2$$

which says that 12 g of carbon reacts with 32 g of oxygen to make 44 g of carbon dioxide. At first glance, this second reaction appears bizarre, but it is also correct. In the first case, we are talking about molecules; in the second case, we are talking about masses. Note that in chemical reactions, moles are not necessarily conserved, whereas mass is always conserved. In the first reaction, one mole of carbon added to one mole of oxygen gives only one mole of CO_2, not two moles. In the second reaction, 12 g of carbon is added to 32 g of oxygen, giving the expected 44 g CO_2. In chemical systems, we will almost always want to talk in molecular terms, but if we are going out to buy chemicals, they will probably be sold by mass.

In classes, we drive home this difference by asking students whether they wanted to be graded on a mole basis or a mass basis. On a mole basis, each student would get one grade. On a mass basis, each student would get their number of points earned divided by their mass. Heavier students would suffer. Somehow, this brings home the point implied by the two equations above. In this book, we will almost always write chemical equations on a molecular basis, that is, on a mole basis; but once we start making economic estimates, we may need to use a mass basis.

Reflection: the sugar glucose $C_6H_{12}O_6$ is metabolized in our bodies to carbon dioxide and water, i.e.

$$C_6H_{12}O_6 + 6O_2 \rightarrow 6CO_2 + 6H_2O$$

This reaction also releases energy. What would this equation be if written in terms of masses?

Energy units. We now turn to the question of the units used for energy. Energy is a quantity both for work and for heat. In this book, we will most frequently describe the amount of energy in the International System of Units or SI units, expressed in joules:

$1\,J = 1\,kg\,m^2/s^2$

Because a joule is often too small to be a useful descriptor of energy, we specify energy by multiples of 1000 J. For

Table 1.3 Common energy units

1 kilojoule (kJ) = 10^3 J
1 megajoule (MJ) = 10^6 J = 0.278 kWh
1 gigajoule (GJ) = 10^9 J
1 terajoule (TJ) = 10^{12} J
1 exajoule (EJ) = 10^{18} J

example, a kilojoule is 1000 J, which is about the heat released when a matchstick head is burned. Other common multiples are given in Table 1.3.

In practice, descriptions of energy use an enormous number of different units, as suggested by Table 1.4. For example, electrical energy is frequently given in kilowatt hours (kWh). Chemical energy is often given in calories. Food energy is also given in Calories, where the upper-case C denotes an energy 1000 times larger than the lower-case c. Mechanical systems are described in horsepower, and air conditioners are often sized in terms of British Thermal Unit (BTU). This complexity has a historical basis and sometimes is more convenient, but it is confusing. Just learn to translate between the units and to focus on the ideas involved. These include ideas like energy can be stored in chemical bonds, used in mechanical devices, and transferred as heat or as work.

Reflection: Having so many energy units may seem confusing. We have thought of this during home remodeling projects. When we are rehanging pictures, everyone seems to have a different ruler, in meters or feet or inches or millimeters, etc. Consider how these differences are resolved.

We want to pause a moment to talk a little about the man for whom the joule is named. James Prescott Joule (1818–1889) was a wealthy craft brewer and a student

Table 1.4 Common conversion for energy units

Name	Symbol	Value	Used in
Joule	J	kg m^2/s^2	All science
Kilojoule	kJ	10^3 J	All science
Megajoule	MJ	10^6 J	Energy policy
Kilowatt hour	kWh	3.6 MJ	Electricity bills
Calorie	cal	4.18 J	Chemistry
Kilocalorie	kcal	10^3 cal	Chemistry
Calorie	Cal	10^3 cal	Nutrition and food
British thermal unit	BTU	1055 J	Furnaces and air conditioners

of John Dalton. Joule's life is a key moment in scientific history. Many scientists older than he were independently wealthy and pursued their scientific interests as a hobby. Younger scientists often became professionals. Joule's life is a watershed in the intellectual history of science, where science starts to be a profession.

Joule was the first to accurately measure how much work is needed to raise the temperature of one cubic centimeter of water one degree Celsius. His measurement in 1843 is within 1% of the value accepted today. But this accuracy was achieved with some personal sacrifice. Joule married late, somewhat to the surprise of his friends. He and his bride then took a honeymoon in the Bernese Oberland, where they met James Clark Maxwell (1831–1879), another remarkable scientist. While hiking near the Eiger, Maxwell saw Joule's carriage approaching. The new Mrs. Joule was in the carriage, but Joule himself was walking behind, carrying his thermometer. He had wondered if being at a higher altitude would change the reading on his thermometer, so he had brought it on the honeymoon. He feared that the jostling in the carriage would corrupt his measurements, so he walked behind the carriage. Later, he and his wife did have three children.

Reflection: A sweet roll is said to have 300 Calories. This is the same as 300 kcal in the more scientific units. Because each kilocalorie is 4184 J, the sweet roll has

$$(4184\, J/kcal) \cdot (300\, kcal) = 1.25 \times 10^6 J$$

or 1.25 MJ. When you try to burn off this sweet roll on a stationary bike, you can take solace in the fact that your body is only about 25% efficient in converting heat to work. The bike says you are producing 100 W of work, that is 100 J/s. Thus, burning off the caloric energy in the sweet roll takes

$$(0.25(1.25 \times 10^6\, J))/(100\, J/s) = 50\text{ min}.$$

"A moment on the lips is a lifetime on the hips." Discuss what can be changed about this discouraging picture.

The second term in the description of energy is power, which is the rate of energy use, that is, the rate of doing work or of heating. The simplest unit of power is the watt:

$$1\,W = 1\,J/s = 1\,kg\,m^2/s^3 = 1\,kg\,m^2/s^3 = 1\,A\,V$$

James Watt (1736–1819), a Scot who dramatically improved the steam engine, was a couple of generations older than Joule. Perhaps because he had some money but did not inherit a brewery, his work was always more practical.

The number of common units for power is smaller than that used to express energy, as shown in Table 1.5. While the ideas of energy and power may not be that difficult,

Table 1.5 Common units for power

Name	Symbol	Value	Used in
Watt	W	1 J/s	All science; equals one amp volt
Horsepower	hp	746 W	Motors

they can be complicated. We will frequently be given a value in one set of units, and we will want to express this in a different set of units. As an example, imagine that we want to know how many food calories (Cal) we will burn to equal one horsepower (hp):

$$1\,\text{hp} = (746\,\text{W/hp})(1\,\text{J/s W})(\text{cal}/4.18\,\text{J})(\text{Cal}/1000\,\text{cal})$$
$$= 178\,\text{cal/s} = 0.178\,\text{Cal/s}$$

Again, there is nothing difficult. Everyone must stagger through the first few conversions to become competent in energy. We now have what we need to begin exploring energy resources and how these can possibly be modified to make more of our society sustainable.

1.3 Why Distinguishing Work from Heat Is Hard (Important and Difficult, but Short)

In the first sections of this chapter, we described how much energy our society is consuming. We have begun to explain why the description of energy is difficult. Part of this difficulty comes from the wide variety of units used in different applications and industries to measure energy, like joules, calories, horsepower hours, and watt seconds. To reduce this difficulty in this book, we will most often express energy in terms of joules and power in terms of watts. Remember, energy per time is power, and one watt equals one joule per second. Everyone can understand this much easily.

However, energy is also complicated because it transfers between systems in two distinct ways. One form of energy is thermal, that is, the energy of heat. The other form of energy is work, that is, the energy of movement. These two forms of energy are like two types of fruit. In discussions in business, we frequently hear the caution, "We can't compare apples and oranges." This has always seemed complete nonsense to us: we do compare apples and oranges every time we go to the supermarket. Apples and oranges look different. But our comparison in the market involves factors like appearance or freshness, or whether or not we need one specific fruit for a recipe. Comparing apples and oranges is relatively easy, less difficult than comparing work energy and heat energy.

The difficulty in comparing energy comes from converting one energy into another. Specifically, work energy can be converted to heat energy with an efficiency of nearly 100% or even more if we use machines like heat pumps to move heat from one location to another. However—and this is the important point—heat energy can *in practice* be converted into work energy but with an efficiency often less than 40%. When we see an efficiency this low, we will immediately be tempted to try to develop better technology to make this conversion more efficient. This turns out to be impossible.

As a tangent, we recognize that we can make something cooler by using heat energy converted to work energy. After all, if we put warm beer in the refrigerator, it gets cold. But we're able to do this only if we add additional work energy. The beer won't get cold if the refrigerator is not plugged in.

What causes this difference? Why is the conversion from work to heat near 100% while the conversion from heat to work is only at best 40%? The difference comes from our experience that heat always moves from hot to cold. Heat does not spontaneously move from cold to hot. In the same sense, solute molecules in a concentrated solution always diffuse into a dilute solution. Molecules do not spontaneously move from dilute to concentrated. We don't expect that a dilute solution like seawater will spontaneously separate into brine and pure water without additional work. We will discuss the reasons for this more thoroughly in Chap. 4; for now, just recognize that it is consistent with experience.

This major difference between thermal energy and work energy is often implicitly ignored or buried in a footnote in large reports from either government or industry. One group will report all the data as heat. For example, the report may give the total energy available as the thermal energy in fossil fuels plus (1/0.4) times the work energy in wind and sun. In the same sense, other investigators may report all the energy as if it were work: adding 40% of the thermal energy of fossil fuels to 100% of the work energy from sun and wind. Other groups simply ignore these problems, adding the heat and work energies together. These choices may make perfect sense for someone working largely in heat energy, or for someone else working largely in work energy, or for someone else who doesn't care about details; but they can be confusing for people who are trying to sort out the differences between them.

1.4 How We Estimate the Future

Reports on energy often contain apocalyptic projections of how we are rapidly depleting our planet's resources. These sources assert we have only a few decades of oil left, or that natural gas prices will go way up, or that we have enough coal for five hundred years. These headlines are based on estimates: no one knows exactly how much coal or natural gas, or uranium is left. Still, these estimates do have some sort of basis. In this section, we want to show how these estimates are made.

Before we quantitatively describe energy resources, we sketch one simple way to estimate the amount of fossil fuels available. This estimation will benefit from a greater knowledge of mathematics, including some calculus and differential equations. Such knowledge is not essential to exploring how we can switch from fossil fuels to sustainable energy resources, so if you are mathematically intimidated, you can skip the material in italics below without compromising your ability to master the rest of the book. Still, if you try, you will learn a little.

Specifically, imagine removing a resource like coal from a coal mine. Normally, we will not know the extent and location of the coal deposits; we still will want to estimate how much more we think we can get in the future. Since this is an estimate, we will not be sure of the answer. Still, we make estimates like these all the time, like guessing our incomes in the next year.

Our only sure present knowledge is the total of past production q in a different mine versus time t, which normally follows a line like that shown in Fig. 1.2. Initially, we will struggle to decide if the coal mine will make money, so little is produced. When we start mining, our production slowly accelerates as we learn what to do. Then q becomes significant, and only then starts to be dependable. Once our mine is reliable, we will

1 Energy Demand

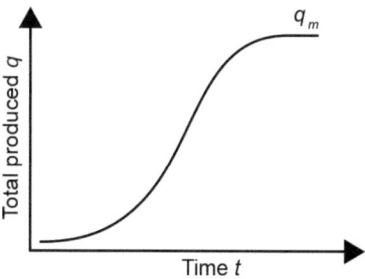

Fig. 1.2 Coal production versus time. Production starts slowly, accelerates, and then slows as the coal mine is exhausted

continue to produce at a steadier rate dq/dt. When the resource begins to be exhausted, the total amount that we produce slows and approaches a maximum limit q_M.

We hope to understand this curve quantitatively so we can plan production. We can do this with different approximations, three of which are discussed here. In the simplest case, the production rate dq/dt will be constant until the resource is exhausted. This behavior, shown in Fig. 1.3a, has two parameters: the slope of the line a—the production rate—which we can regulate; and the maximum amount of resource q_M, which we may be able to guess from similar mines. This simplest approach is normally too primitive to have practical value.

The second model recognizes that we are most concerned when the production rate dq/dt begins to decline, as shown in Fig. 1.3b. We assume that the amount produced is proportional to the amount remaining:

$$dq/dt = (q_M - q)/a$$

where a is a time typical of this type of mine. This is subject to the initial condition that the amount produced is zero at some starting time t_0:

$$t = t_0, q = 0$$

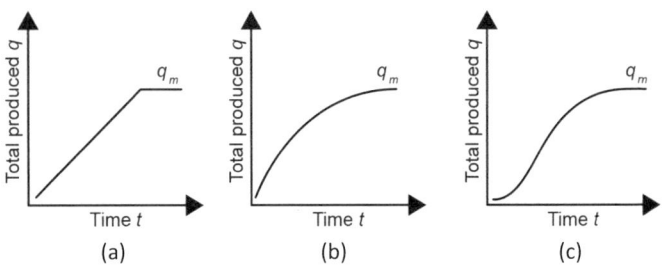

Fig. 1.3 Three approximate estimates of the amount produced. In **a**, the rate is constant till exhaustion; in **b**, it rises exponentially to a limit; and in **c**, it follows a logistic curve

Integrating we find that q varies with time according to the relation:

$$q/q_M = 1 - e^{-(t-t_0)/a}$$

We now have three parameters: the total amount q_M, estimated from similar mines; the time t_o, when we begin production; and the depletion time a. We can now estimate how the productivity of the mine decays.

The third model, the logistic approximation, asserts that the amount produced by a resource will behave like that in Fig. 1.3c. As before, q is the total amount produced, q_M is the maximum which can be produced, and a is a characteristic time. Such behavior is closest to what we expect, though our data may not be good enough to make us confident of the values of our three parameters. This model postulates a differential equation:

$$dq/dt = q(1-q/q_M)/a$$

subject to an initial condition:

$$t = 0, q = q_0$$

1 Energy Demand

where q_0 is an exceedingly small initial mass. As in the case of the exponential model, this gives the amount produced q as a function of time and three variables: q_0, q_M, and the production time a. This expression, which is widely used, often gives a good description of the reservoir behavior. Integrating and using the initial condition gives:

$$q = q_0 e^{t/a} / (1 + (q_0/q_M)(e^{t/a} - 1))$$

The equation has two interesting limits. First, when the time is small, it predicts that q will equal $q_0(1 + (1 - q_0/q_M)(t/a))$, which is linear with time. Second, it predicts that when time is very large, q will approach the limit q_M.

While the third model often represents our experience, the quantity q_0 does not have a strong physical significance. As a result, those working on this topic often replace this parameter with an alternative, like the time for producing half of the resource. We will try to fit our production data to this equation, and then use the equation to predict both current and eventual production. We will do this for all the mines we know about and thus estimate how much we have.

This estimation is speculative, a reflection of our data and our experience. For the purposes of this book, we want to emphasize how much estimation is involved, that is, how much curve fitting of production is needed for resource estimation. We find it impressive that estimates of resources can be as effective as they are, given the enormous ignorance of the actual physical structure of the resources before the resources are removed.

Before leaving this mathematically based development, we want to mention that parameters like production rate depend not only on technical factors but also on economic factors. In a free market, production is a balance between supply and demand. And in some cases, the supply or the demand can radically change. A tax on carbon would slow down demand; a drop in production rate could follow. Technical advances in *mining could make it cheaper to produce more of the coal and thereby increase supply. Such changes are constantly happening, not only in fossil exploration and production, but also in other forms of energy generation. We shall discuss these in further detail in the next chapters.*

1.5 Conclusions

Energy is key to our society. It is how we went from being a race of hunter-gatherers to partly controlling our environment. In the past—and even now—our society has depended on generating that energy by digging up things from the ground and burning them. Because carbon is abundant in organic matter and oxygen is plentiful in the atmosphere, we first burned carbon-based wood and later shifted towards burning energy-dense carbon-based fossil fuels deposited over millions of years. Today, fossil fuels provide over 80% of the world's energy. Producing this energy results in the formation of carbon dioxide, that is, CO_2. Today, we release this CO_2 into the atmosphere directly, and this is primarily responsible for climate change.

We want to explore the possibility that we can convert our energy use from fossil fuels to sustainable resources. These resources depend on solar energy in all its forms, including sunlight and wind. This conversion will include building a new infrastructure, like that now used to distribute fossil fuels. For example, we may need to augment or replace gas stations with electrical charging stations. Doing this replacement will require money and will take decades, depending on how society decides to act. We took about 50 years to establish our network of gas stations. Exactly how we could undergo this energy transition is explored in the next chapters.

References

International Energy Agency. 2025. *World Energy Outlook 2025*.

The Energy Institute. 2025. *Statistical Review of World Energy*, 74th ed.

U.S. Department of Energy. 2025. *Lighting Choices to Save You Money*. Accessed 1 Dec 2025. https://www.energy.gov/energysaver/lighting-choices-save-you-money

U.S. Energy Information Administration. 2025a. *U.S. Energy Consumption by Source and Sector 2024*. Accessed 1 Dec. 2025. https://www.eia.gov/totalenergy/data/monthly/pdf/flow/total_energy_spaghettichart_2024.pdf

U.S. Energy Information Administration. 2025b. *U.S. Energy-Related Carbon Dioxide Emissions 2024*.

2

Unsustainable Energy: Coal

One big contribution to climate change comes from the use of coal. Coal is a rock that burns. Every year, burning coal globally releases about 178 EJ of thermal energy and emits over 15 Gt of CO_2 (International Energy Agency 2025). In total, this heat could be converted to around 70 EJ of work using machines like power plants or train engines.

Coal combustion has a long history. In the time of Julius Caesar, those living near Genoa burned surface lumps of coal for heat and cooking. In China, Marco Polo was impressed by the Chinese, who took three baths per week with water heated by coal. British shaft mining of coal began about 1200 CE. Coal-fired steam engines replaced waterwheels by 1700, starting and sustaining the Industrial Revolution.

Coal is a hydrocarbon made from fossilized organisms, which themselves are mostly carbon, hydrogen, and oxygen. We use these fossilized hydrocarbons as fuels in the form of coal, oil, and natural gas. Since the industrial revolution and even today, the high energy density, abundance, and low cost of fossil fuels have provided most of society's energy. Currently, fossil fuels provide well over 80% of the energy we humans consume, a number that has not changed much for decades. Virtually all of this fossil fuel is combusted, and the resulting CO_2 is released into the atmosphere. This needs to change if we are to mitigate climate change.

Reflection: The Industrial Revolution is often said to have sparked interest in coal as an energy source. Many assert that this was because early industrialists ran out of wind power. Do you believe this?

2.1 What Coal Is

Coal is not a single compound but a mixture that has a carbon-to-hydrogen mole ratio between 0.3 and 0.7, i.e., between $CH_{0.3}$ and $CH_{0.7}$. The energy density per mass of coal is high, between 15 and 33 MJ/kg. The exact value depends on the type of coal, its water content, and the presence of inert compounds, which will produce ash. For our discussion here, we assume the chemical formula of coal is about $CH_{0.5}$ and has an energy density averaging 30 MJ/kg. Coal combustion produces heat energy from the reaction, producing CO_2 and water:

$$CH_{0.5} + 1.125\ O_2 \rightarrow CO_2 + 0.25\ H_2O$$

Both carbon and hydrogen in this reaction come from coal, while oxygen comes from air.

The actual chemical compounds in coal and other fossil fuels vary widely. Some of the compounds are low molecular weight liquids, more familiar as compounds in gasoline. These include small amounts of aliphatic chains, like n-heptane (C_7H_{16}):

Aromatic compounds like toluene (C_7H_8) may also occur:

Combustion of each of these compounds contributes to the overall average heat of combustion of 30 MJ/kg. The physical properties of compounds like these are different because their chemical structures are different. For example, n-heptane has an octane rating of 0, and toluene's octane rating is 114. Structure matters.

These two drawings use different ways of representing the two chemical compounds. They have the same number of carbon atoms but different numbers of hydrogen atoms. For heptane, we detail the location of all the carbon atoms and of all the hydrogen atoms, but for toluene, we abbreviate the carbon atoms simply as the intersection of two lines or the endpoint of a single line. We don't show hydrogen atoms explicitly, but must infer their number and location from the fact that each carbon atom chemically bonds to several adjacent atoms while each hydrogen atom bonds only to one other adjacent atom. Lines between carbons are the chemical bonds made with electrons, and the number of lines shows the number of electron pairs in the bonds. As a third example, naphthalene ($C_{10}H_8$), which is the key component in those round, white, smelly, old-fashioned mothballs, has the structure:

Again, when we want to estimate how much energy we get from burning these compounds, we need to know their structure. But in practice, the structures in coal are even more complex. For example, one fraction of coal has a structure like the following:

To estimate the properties of a fragment like this, we must resort to experiments.

Moreover, this piece of coal could also contain impurities, including sulfur (S), silica (Si), and mercury (Hg). Silica and mercury are not shown in the figure because often they're not chemically attached to these hydrocarbons. Instead, they're attached to other molecules that are part of the coal mixture. We really must depend on experiments to determine the properties of the specific coal in question.

The presence of impurities in coal can have significant environmental consequences. Mercury is an excellent example. It is toxic. Mercury exists in coal only at concentrations on the order of 0.1 parts per million (ppm), but it is still present in the gaseous emissions that come from burning coal. In coal-fired power plants, emissions are mostly controlled via environmental control technologies that remove particulate matter, sulfur oxides, nitrogen oxides, and mercury from flue gas. Particular matter still released into the atmosphere is linked with lung ailments, sulfur and nitrogen oxides result in acid rain. Mercury released into the atmosphere can be concentrated by living plants and animals. This effect, called bioaccumulation, means wild salmon can potentially contain higher concentrations of mercury than faster-growing farmed salmon. When we establish environmental policy, we need to be careful of impurities that occur because of bioaccumulation.

2.2 Where Coal Is

Coal is formed from the decay of wetlands under varying temperatures and pressure over millions of years. This decay is often approximated by the sequence of forms of coal:

vegetable matter → peat → lignite → subbituminous → bituminous → anthracite

During this slow process, the temperature is more important than the pressure, although both are involved. As we move from left to right, the material has less water and more carbon.

When we were children in the 1950s and 60s, our houses were heated by anthracite, the coal then found especially in northeast Pennsylvania. Unlike anthracite, bituminous coal was hard to burn to heat a house, but basic to the manufacture of steel. Once, when we came home on a grey rainy day, we built an anthracite coal fire to warm the empty house and to cheer up. The fire ignited the chimney in our house, and we learned that chimney fires are hard to put out. At other times, we didn't have running hot water in our homes, and our parents used bituminous coal to heat water buckets for taking warm showers. Some communities throughout history have also used peat as a cheap fuel, though environmental concerns are limiting its use today.

There is an enormous amount of coal left on the planet. If all the proven coal deposits were used at the current rate, there is still at least a 150-year global supply. With such a large supply already known, there is little incentive to find additional ones. The United States has 22% of the world's reserves, enough to supply 500 years at current domestic consumption rates. Other major sources are Russia (14%), Australia (13%), China (13%), and India (10%) (U.S. Energy Information Administration 2023). A century ago, most coal was mined in shafts: there were a million coal miners working underground in the United Kingdom and 800,000 in the United States. Today, British mine shafts are closed, and only 50,000 miners work below ground in the United States. Now, almost all coal mining is on the surface. The chief producer in the United States is the state of Wyoming, followed by West Virginia. But coal in the US is in decline, as shown by the data in Fig. 2.1 (Data from Boston University 2023). Coal in the US was surpassed by oil in 1950.

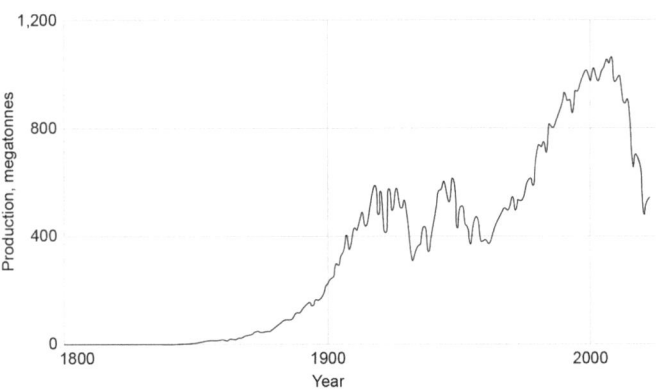

Fig. 2.1 Coal production in the U.S. The production of coal in the United States is in sharp decline (Data from Boston University 2023)

Because coal has been cheap and plentiful, most countries, including the United States, have used coal to generate energy for manufacturing. As a country's economy develops, jobs gravitate more towards service-based sectors, and manufacturing shifts to less-developed countries. Over time, the less-developed countries also move towards service-based sectors, and manufacturing shifts once again.

This manufacturing shift has occurred not only for coal but for other fossil fuels too, across industries like textiles, steel, chemicals, cars, and electronics. While demand for products like iPhones comes from developed countries, the phones themselves are manufactured in developing countries like India and China, which rely on plentiful, low-cost coal. This trend has caused coal use to drop in the most developed countries and to increase in the developing ones, as shown in Fig. 2.2 (The Energy Institute 2025). But the overall total global coal consumption continues to rise and is now at record levels. Whether total coal consumption increases in the coming years will depend on the cost of energy from other sources and on carbon mitigation policies around the world.

2.3 What Coal Is Used for

In the mid-1900s, coal in the U.S. was widely used for transportation, heating, industry, and electric power. Today, it's overwhelmingly used to generate electricity, as shown in Fig. 2.3. (U.S. Energy Information Administration 2023b). Other uses are to make coke, a key to steel production, and to make cement, a major part of buildings. Each of these processes emits CO_2 and, hence are factors in global warming.

Over 90% of the coal consumed in the U.S. is used to make electricity in a power plant, as shown in Fig. 2.4

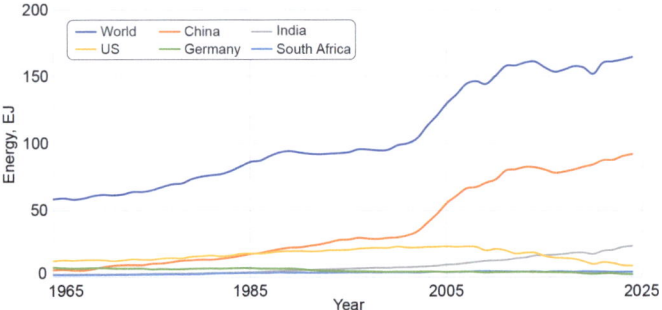

Fig. 2.2 Coal Consumption by Several Large Countries. China's major role in coal is complemented by it having the world's largest increase in solar-generated electricity (Data from The Energy Institute 2025)

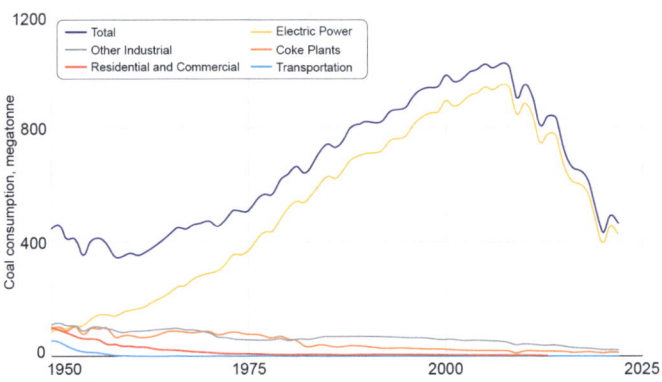

Fig. 2.3 Major Uses of Coal in the U.S. Coal is largely used to make electricity. Other uses are steel and cement (U.S. Energy Information Administration 2023b)

(TVA 2025). The coal is burned in air to generate heat, which then boils water to make steam. The steam is expanded in a turbine, where blades rotate a shaft. The rotating shaft is also attached to a generator, where electromagnets spin in proximity to coiled wires to generate electricity. This electricity is carried by wires to a transformer and then to the electrical grid. The steam exiting the turbine is cooled to make liquid water, which is recycled back to the boiler. This cooling can be typically done by more water, or sometimes air, but is a necessary part of any heat-to-work machine, as will become evident in Chap. 4.

Thus, a power plant is a machine that converts chemical energy in the fuel into thermal energy of the steam into kinetic energy of the spinning turbine blades into electrical energy in the generator. Modern coal-fired power plants are about 40% efficient overall, meaning for every 100 J of thermal energy entering the plant, about 40 J of work energy are generated as electricity. Many experts are working to improve power plant efficiency, but as will be explained in Chap. 4, there is a maximum possible efficiency of any heat-to-work machine.

As shown in Fig. 2.3, coal is also used for steel and cement. Making steel requires making coke, produced by the destructive distillation of bituminous coal at above

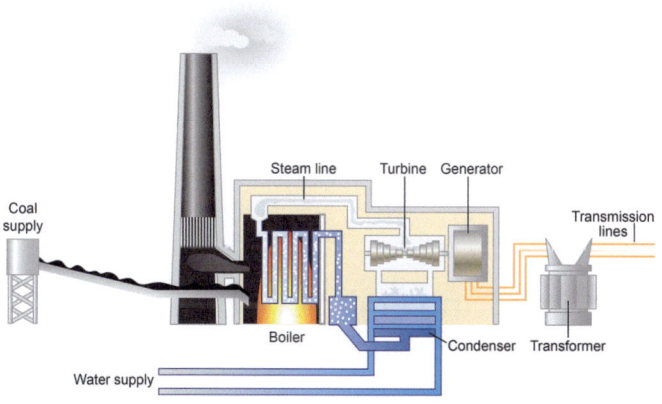

Fig. 2.4 How coal is burned to produce electricity. Coal is burned to make steam, which rotates the blades of a turbine. The turbine turns a generator to make electricity (TVA 2025)

1000 °C to produce a nearly pure form of highly porous carbon called coke. Typically, 1500 kg of coal makes 1000 kg of coke. The coke then reacts with an iron ore like magnetite (Fe_2O_3) to make iron (Fe):

$$2Fe_2O_3 + 3C \rightarrow 4Fe + 3CO_2$$

One tonne of coke makes around 1.6 tonnes of iron; the production of iron and steel results in 7% of global CO_2 emissions (International Energy Agency 2020, 2025).

Coal is also used indirectly for producing cement. In this reaction, calcium carbonate ($CaCO_3$)—that is, limestone—is heated to produce calcium oxide (CaO) and CO_2:

$$CaCO_3 \rightarrow CaO + CO_2$$

In addition to the CO_2 produced by this reaction, heating the $CaCO_3$ uses more coal. The total CO_2 produced by reaction and by heating in the manufacture of cement is 6% of total global CO_2 emissions (International Energy Agency 2025).

Reflection: The early chemical business was closely connected with the coal business, especially through "coal liquids" made from coke manufacture. What were these liquids?

2.4 Reducing Emissions by Carbon Capture and Storage (CCS)

All fossil fuels, and especially coal, produce large amounts of CO_2 in the flue gas, which is exhausted. While most agree that CO_2 is a major world problem, many believe this problem can be mitigated solely by switching to renewables. However, renewables provide energy intermittently, solar energy when the sun shines and wind energy when the wind blows, as further discussed in Chap. 6. Most energy forecast models show that we are going to keep burning fossil fuels even as we switch to renewables to meet society's demand for energy. However, these forecasts are controversial. We do not present a conclusion here, but remind you that in this book, we are discussing what we could do, not what we should do.

To reduce the effects of burning coal and other fossil fuels while we switch, many urge carbon capture and storage (CCS), which we will explore in more detail in Chap. 8. For now, we note that there are normally four steps in this process:

(1) Capture—separation of the CO_2 from other gases;
(2) Compression—compression of the gaseous CO_2 to make it liquid-like;
(3) Transportation—moving the CO_2 from the capture site to the storage site;
(4) Storage—permanent underground burial of the CO_2.

CCS can also be applied not to just CO_2 from fossil fuels, but also to other gases that contain CO_2.

2.5 Biomass, a Tangent

So far, we have been discussing coal; in the next chapter, we will discuss oil and natural gas. Tangentially, we want to briefly discuss biomass. Biomass is a form of renewable energy that comes from plants that the earth can naturally replenish. We use this term for convenience only, recognizing that the sun is ultimately the energy source for most forms of energy we use. The sun provides not just solar energy, but it also moves the wind to provide wind energy, drives the snow and rain that eventually provide hydroelectric power, powers photosynthesis to make biomass from CO_2 and water, and it even decays organic matter into fossil fuels over millions of years.

Renewable energy encourages the idea that we grow our own fuel. We harvest the sun's energy not as electricity but as a crop that we can use as fuel. The crop most often mentioned is corn, grown on farms that already exist. Corn can be fermented to produce ethanol, a motor fuel. Other crops include wood chips that can be burned much like coal or natural gas in a power plant. In this section, we want to explore to what extent ideas like this make sense.

Biomass is usually neither a solid nor a pure liquid, but a heterogeneous mixture of the two. It is currently responsible for less than 5% of total energy use in the U.S. and about twice as much in the world. About 45% of energy from biomass is used for industry including wood products and paper; about 36% is used for transportation fuel; about 9% is used for residential use such as firewood; about 7% is used to generate electricity in small power plants; and about 4% is used commercially including biogas produced by waste landfills (U.S. Energy Information Administration 2023a). While burning biomass produces CO_2, a similar amount of CO_2 is consumed in growing the crop. This crop does require energy for tilling and fertilizing.

Biomass combustion typically yields an energy per kilogram about half that of other, more conventional fossil fuels, as shown in Table 2.1. The first column in this table is the fuel itself; the second is its empirical formula; and the third is its heat of combustion. Biodiesel, a renewable fuel popular in Europe, has a structure roughly like.

Table 2.1 Heats of combustion of biomass

Fuel	Empirical formula	Heat of combustion, MJ/kg
Coal	$CH_{0.8}$	32
Oil	CH_2	45
Ethanol	C_2H_5OH	30
Biodiesel	CH_2O	37
Wood	$C_6H_{12}O_6$	15
Green wood chips	–	12
Corn stover	–	17
Switch grass	–	22
Bagasse	–	5

The amount of energy produced from biomass seems unlikely to be a major factor in finding sustainable energy

As Table 2.1 shows, coal has a heat of combustion of about 32 MJ per kilogram. In contrast, firewood has around 15 MJ per kilogram, a value very dependent on how much the firewood is dried. Bagasse, a byproduct of manufacturing sugar, is probably the largest single biomass now used as an energy source; its heat of combustion is smaller than firewood because it contains so much water. Other materials, like green wood chips, corn stover, and switch grass, are mixtures of cellulose, hemicellulose, and lignin; their heating value depends strongly on the amount of water they contain. None is outstanding.

Reflection: These biomass results are highly dependent on the amount of water they contain. Discuss how you can remove water cheaply. Moreover, the water removal gets harder and harder as removal proceeds. This is sometimes called removal of "free water" (easy) and "bound water" (hard). Discuss

what you think this means, and how these factors affect the utility of biomass.

While at first biomass may seem an attractive alternative to fossil fuels, the land needed for its production is huge. Transportation of biomass can be costly relative to its energy content. Thus, biomass-fired power plants tend to be small, relying on transporting biomass from nearby regions. Sometimes biomass production may even displace food production because they both use land and water. The scale of biomass production is currently small and may be hard to increase dramatically. To examine this in more detail, we note that the U.S. energy use is currently 20 million barrels of oil per day, which has the energy content of about 30 million barrels of ethanol per day. At present, making ethanol uses five billion bushels of corn per year. Because a bushel of corn can produce 2.8 gallons of ethanol, and there are 42 gallons in a petrochemical barrel, this corresponds to:

$$(5 \times 10^9 \text{ bushels/year})(\text{year}/365 \text{ days})(2.8 \text{ gal/bushel})(\text{bbl}/42\text{gal}) = 10^6 \text{bbls ethanol/day}$$

The million barrels of ethanol made daily now is less than 5% of our energy use. Making ethanol already takes about a third of the total land devoted to corn. To meet our energy demand just for oil, we would need to increase the amount of land devoted to corn by over 30 times. This seems unlikely.

Reflection: "Rocket stoves", developed by the Aprovecho Research Center and others, are widely used in underdeveloped countries and sold in sporting goods stores. They reduce the energy for cooking by 70%. Do they really work? Would these stoves mitigate the energy crisis?

2.6 Conclusions

From the material above, we see that coal is a reliable, mature technology for obtaining heat energy. Coal is abundant and easily stored. It is not especially attractive when the total societal cost is included, and because competing technologies—especially solar and wind energy—have developed so rapidly. Coal causes pollution of both greenhouse gases like CO_2 and traces like mercury accumulated by living organisms, including fish. Coal also produces many particulates smaller than 2.5 μm, the sizes most implicated in pulmonary disease. Regardless, coal is going to remain an important component of energy generation, especially in developing countries. It is a baseline with which other energy-producing technologies will be compared.

References

International Energy Agency. 2020. *Iron and Steel Technology Roadmap*.
International Energy Agency, 2023. *Cement*. Accessed 1 Dec 2025. https://www.iea.org/energy-system/industry/cement.
International Energy Agency, 2025. *World Energy Outlook 2025*.
Institute for Global Sustainability, Boston University. 2023. *The History of Coal Production in the United States*. Accessed 1 Dec 2025. https://visualizingenergy.org/the-history-of-coal-production-in-the-united-states/.
The Energy Institute. 2025. *Statistical Review of World Energy*, 74th ed.

U.S. Energy Information Administration. 2023. *Coal Reserves*. Accessed 1 Dec 2025. https://www.eia.gov/international/data/world.

U.S. Energy Information Administration. 2023a. *Biomass Explained*. Accessed 1 Dec 2025. https://www.eia.gov/energyexplained/coal/use-of-coal.php.

U.S. Energy Information Administration. 2023b. *Coal Explained: Use of Coal*. Accessed 1 Dec 2025. https://www.eia.gov/energyexplained/coal/use-of-coal.php.

Wikipedia. 2025. *Tennessee Valley Authority (TVA)*. Accessed 1 Dec. 2025. https://commons.wikimedia.org/wiki/File:Coal_fired_power_plant_diagram.svg

3

Unsustainable Energy: Oil and Gas

As part of our jobs, we sometimes took lecture tours to talk at other universities and to groups of professional engineers. These talks took us to centers of the fossil fuel industry like Midland, Texas. Midland is in the Permian Basin, the largest single oil formation in the world. On one trip, we arrived in Midland late in the evening after a talk at Texas A&M; we were not due for the next talk until 5:00 pm the next day. When we tried to do other work, we got stalled. The clerk at the motel told us there was a museum of oilfield equipment nearby, so we decided to visit.

While the museum today has a building, at the time, it was a huge open field with obsolete equipment dumped without documentation. The equipment was huge, lurking like dinosaurs on the grassy plain. Delightfully, these dinosaurs had largely been manufactured in Pennsylvania, not in Texas, and they were made not of metal but of oak beams, bent after steaming. In the 1920s and 1930s, oil producers used this equipment to produce the crude oil that supplied energy for the entire economy, especially the growing auto industry.

The American oil industry had begun in Titusville, Pennsylvania, in 1859, and then exploded in places like Oil City, Pennsylvania. When we started teaching in Pittsburgh in the late 1960s, we had students from Oil City for whom the oil rush was as important as the gold rush in California. The students had lived in huge, poorly maintained houses built by oil millionaires. While the Pennsylvania oil boom was over in the late nineteenth century, Pennsylvania crude continued to be highly valued because it had a large fraction of "lube stock," that is, of crude oil that could be made into lubricants. Quaker State and Pennzoil had survived because their Pennsylvania crude had a high fraction of lube stock.

The entire oil industry boomed, and it never looked back. Global crude oil production now is at historically high levels, around 85 million barrels per day. The United States is now producing around 13 million barrels of crude oil per day, the most of any country ever. Each of these barrels contains 42 gallons, a standard chosen in 1888 by a few producers in western Pennsylvania. Paradoxically, each barrel of crude produces about 1.05 barrels of petroleum products, largely gasoline, because the density of petroleum products is less than the density of the crude oil. Additional hydrocarbons from natural gas liquids and biofuels brings the total global oil production to 100 million barrels per day and the U.S. to 20 million barrels per day, about the same as its consumption.

While the oil industry is a leviathan, it is recent. Ninety percent of the oil ever used has been produced since 1957. That year, almost half of the industrial firms listed in the Dow Jones Industrial Average were oil or chemical companies; today, two are. This is the aging industry with which the world is now grappling.

Natural gas and oil are often found together in the earth. They result from organic material deposited in ancient seas, typically 100 million years old. Coal is formed in similar deposits not on seabeds, but on land. Under heat and pressure, this organic material decayed into fossil fuels. These fuels all contain carbon because life forms on earth are carbon based.

The hydrocarbons in oil and natural gas are a complex mixture, ranging from molecules that contain only one carbon atom to ones that contain hundreds of carbon atoms. Under atmospheric conditions, fluid hydrocarbons that contain one to four carbon atoms are called "light". They are typically gaseous. Those with five or more atoms are called "heavy". They are typically liquid. The gaseous mixture is natural gas, while the liquid mixture is crude oil.

3.1 What Natural Gas Is

Natural gas is typically 95% methane (CH_4), 4% ethane (C_2H_6), < 1% propane (C_3H_8), and even less butane (C_4H_{10}). It is an extremely important commodity for generating energy, for making both heat and work. This energy is usually released by burning natural gas in air, where the carbon becomes CO_2 and the hydrogen becomes water. For the same amount of energy release, molecules that have a smaller carbon-to-hydrogen ratio release less CO_2. Burning a given mass of natural gas, therefore, generates less CO_2 than does burning the same mass of oil or coal. Because it's a gas, it also does not contain as many contaminants like ash, mercury, and sulfur found in coal and oil. It does contain hydrogen sulfide (H_2S). Nonetheless, natural gas is sometimes called a "cleaner" fuel, but it still generates significant amounts of CO_2.

Because of fracking, discussed later in this chapter, natural gas production in the U.S. has risen sharply over the past decades. Gas prices fell equally sharply, making gas cheaper to use than other forms of energy. Electricity generated from natural gas is currently cheaper than coal. It provides more than 40% of the electricity generated in the U.S., more than any other large-scale option. The last coal power plant in the U.S. was built in 2013, and many power companies have announced they are phasing out more coal-fired power plants. Power plants switching from coal to "cleaner" natural gas is the single largest reason why annual U.S. CO_2 emissions over the past 15 years have dropped 20%. The second reason is increased usage of renewable energy, a topic we shall begin to cover in Chap. 5.

Natural gas is primarily used for generating heat, providing about 25% of global energy consumed. It's used widely for heating homes, heating water, and cooking food. It is widely used in industry for boilers, furnaces, and generating electricity. Generating electricity using natural gas is similar to generating electricity using coal, which we discussed in Chap. 2.

Still, there are key differences, as shown in Fig. 3.1. Natural gas is burned in combustion gas turbines to generate CO_2, water vapor, and heat. These hot gases then expand through the turbine blades attached to a rotating shaft, which in turn is attached to generators that produce electricity. The turbine's exhaust gas is still hot and can be used to boil water into steam in a heat recovery steam generator or HRSG, sometimes pronounced "hersig". This steam is then sent to steam turbines that produce additional electricity. The exiting steam is condensed to water and recycled back to the HRSG. Because the electricity is generated using both gas turbines and steam turbines, this type of power plant is called a natural gas combined cycle (NGCC) power plant. Its efficiency of converting

3 Unsustainable Energy: Oil and Gas

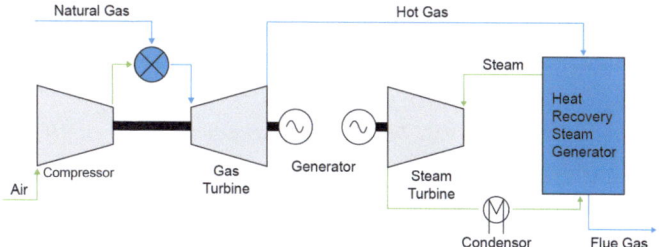

Fig. 3.1 Natural Gas Combined Cycle Power Plant. Gas turbines and steam turbines are used to make electric power from natural gas

heat energy to work energy is approaching 60%, one of the highest energy efficiency machines ever developed. This high efficiency, coupled with a low carbon-to-hydrogen ratio, results in CO_2 emissions per unit power about half that of a coal-fired power plant.

There's a debate on whether these lower CO_2 emissions from natural gas are beneficial to climate change. Production of natural gas from underground reservoirs and its above-ground processing inevitably leads to some methane leaking into the atmosphere, often called "fugitive emissions." When averaged over a hundred years, methane is 30 times more potent in generating atmospheric heating than CO_2. When averaged over 20 years, it's 85 times more. Fugitive emissions of methane can negate the "cleaner" benefits of natural gas, potentially making it environmentally worse than coal. The debate is not completely settled, mostly because of the uncertainty in quantifying fugitive emissions.

Because oil and gas are often found together, fracking for increased gas production also leads to increased oil production. Unlike natural gas, oil is rarely used to generate electric power. Instead, oil is used mostly for energy for transportation, as discussed next.

Reflection: Why is natural gas felt to be less environmentally abusive than coal? Is it?

Reflection: Advocates for gas say it is a greener fuel. Coal producers retort that because methane warms the earth 30–85 times more than CO_2, fugitive methane emissions make gas as least as bad. Who is right?

3.2 What Crude Oil Is

Crude oil and natural gas are both fuels and feedstocks for making chemicals. The oil, held in porous rock, contains hundreds of chemical compounds. The original uses of crude oil centered on medicines and fuel for lamps. Lamp fuel made from whale oil peaked about 1845. Whale oil was largely triglycerides, that is, glycerol attached to different fatty acids. Whale oil was first replaced by camphine, a mixture of turpentine and alcohol similar to that sold today as lamp fuel. Odorless camphine, a chemical mixture, is sometimes confused with smelly camphene, a specific natural product made from pinene. Camphine, in turn, was replaced by kerosene, a mixture containing 9–16 carbon atoms per molecule and made from crude oil.

Crude oil, the source of both kerosene and gasoline, is a complicated mixture of alkanes and aromatics. The alkanes include straight chains of 8–20 carbon atoms. Normal octane (n-C_8H_{18}) is an example. The branched octane 2,2,4-trimethylpentane is the standard for motor fuel, rated "100 octane." n-Heptane (n-C_6H_{14}) is zero octane. Other major components of crude oil are aromatics, which do not have so many carbons with single bonds. Benzene (C_6H_6) and toluene (C_7H_8) are examples. Pure toluene has only seven carbons but an octane rating of 115. Significant amounts of toluene are used in airplane fuel.

Reflection: Oil is a complicated chemical mixture but is sometimes idealized as four species: aliphatics, aromatics, branched, and alcohols. Discuss what these terms mean.

Crude oil is produced by digging a hole in the ground, finding a reservoir of oil, and pumping the oil out. In some cases, the oil is already under pressure and will flow up naturally. In many more cases, the natural pressure is insufficient, and the crude oil is pushed out by injections of CO_2 and steam or detergent-loaded water, as shown in Fig. 3.2 (U.S. Department of Energy 2008). In this process, called CO_2-enhanced oil recovery (EOR), some of the CO_2 is effectively permanently stored underground as it displaces the oil. A typical EOR operation stores 0.3–0.6 tonne CO_2 for each barrel of oil produced, or roughly one tonne CO_2 per two barrels of oil produced. That same barrel of oil releases about 0.4 tonne CO_2 when combusted, with an additional 0.1 tonne CO_2 emissions due to production, processing, and transportation.

This suggests oil produced with EOR could be "carbon neutral" or even "carbon negative", depending on the amount of CO_2 stored underground relative to that released from oil combustion and processing. The CO_2 most often comes from a naturally occurring underground reservoir. But it can also come from a carbon capture process attached to a power plant or an industrial site that combusts fossil fuels or biomass. Whether or not a particular CO_2-EOR process is deemed to be carbon positive, neutral, or negative depends on the origin of the CO_2, characteristics of the oil well, and the boundaries used for carbon accounting. Still in 2022, to reduce the production of fossil fuels, California banned using CO_2 for EOR if it was captured from any carbon capture process. Due to technological advances such as EOR, horizontal drilling, and fracking discussed in the next section, the United States is now the world's largest oil producer.

Reflection: How much of the oil in a typical formation is recovered with conventional recovery? Discuss the ways oil producers increase this recovery.

Crude oil in the U.S. was first produced in quantity by Edwin L. Drake in Titusville, Pennsylvania. After a brief boom, the oil industry as a whole was reorganized as a monopoly by J. D. Rockefeller. Based in Cleveand,

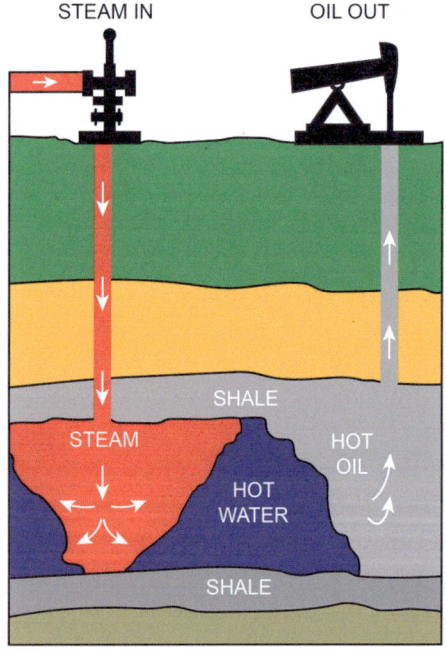

Fig. 3.2 Producing Crude Oil with Steam Injection. The steam reduces oil viscosity, facilitating flow (U.S. Department of Energy 2008)

Rockefeller was a hardworking, religious, and crafty businessman. His monopoly, the Standard Oil Corporation, led to the passage in 1890 of the Sherman Antitrust Act, used to break up Standard Oil in 1911. In inflation-adjusted dollars, he was the richest man the world has ever known. His fortune was estimated as 2% of the national wealth, 20 times larger than today's richest man.

The oil industry left Pennsylvania for Texas, where it was led by Columbus Marion "Dad" Joiner. The first major well was Spindletop, a gusher which erupted in 1901. Joiner thus found the East Texas field centered in Rusk County. While Texas still produces a million barrels of oil per day, its boom only lasted from about 1930–1960. Then oil obtained abroad, especially from the Middle East, became more important, as shown in Fig. 3.3 (Data from Xu and Bell-Hammer 2023). The advantages of Middle East oil were first recognized by Winston Churchill, who redrew the map so that Britain could control oil from Iraq. Other key areas include Venezuela and Mexico, responsible for a large fraction of Central and South American oil; Russia and the former Soviet Union, which includes reserves around Baku; and China, both in the northwest and offshore in the southeast.

Reflection: Rockefeller is usually believed to be a very astute businessman in assembling Standard Oil. But at the time he did so, the lighting business, which had dominated petroleum use, was threatened by the development of electricity. The gasoline market for automobiles was not developed. Do you think Rockefeller was smart or lucky?

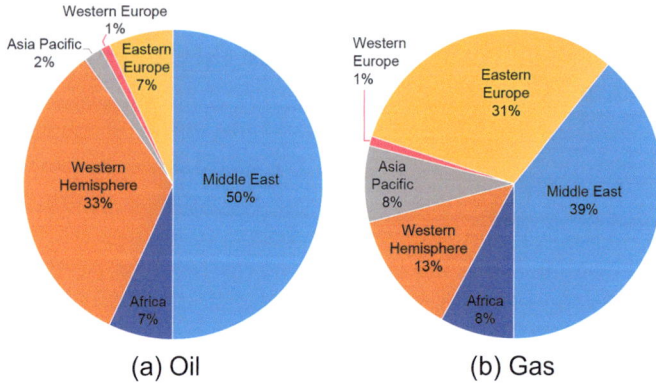

Fig. 3.3 Where oil and gas are located. Most proven reserves of oil and gas are in the Middle East (Xu and Bell-Hammer 2023)

3.3 How Much Oil and Gas Are Left

The greatest question for the oil and gas industry is how much fuel is left and how to extract it at prices that people will pay. Today, the world still has proven reserves of approximately 1.75 trillion barrels of oil and 7.5 quadrillion cubic feet of gas (Xu and Bell-Hammer 2023). Global oil consumption is about 100 million barrels per day, or about 200 EJ per year, while global gas consumption is about 400 billion cubic feet per day, or about 150 EJ per year (Statistical Review of World Energy 2024). Even at these record-breaking consumption levels, which show no signs of slowing, proven reserves for oil and gas are about 50 years each. Does this mean the world will run out of oil and gas in 50 years?

Simply put, no: we are not about to run out of oil. The term "proven reserves" is often misunderstood. It means that there's a reasonable probability that at least 90% of an oil and gas reservoir is profitably extractable at current prices. As consumption diminishes proven reserves, oil and gas companies find new reservoirs, develop new technologies to lower the cost of production, and thus increase proven reserves. Over time, such market forces create a balance between increased consumption and increased production, which is about 30–50 years. Indeed, there is little incentive for companies to find new reservoirs or to invest in technologies that would be useful beyond 30–50 years. Even back when we worked in the oil industry 35 years ago, the world had proven oil reserves of 30–50 years, just like today. In reality, there are scores of oil and gas reservoirs inside the earth, waiting to become "proven." No doubt the earth is finite and will eventually run out of fossil fuels at some point, but that will not be for many decades, if not centuries. Fossil fuels will certainly not disappear in the time that we have left to mitigate climate change.

This question, "How much is left?" was explored by Marion King Hubbert. Born on a farm in West Texas, Hubbert began studying at Weatherford College. After two years, the college president told Hubbert to move to the University of Chicago, where he would have a full scholarship. He graduated in 1926 with a bachelor's degree in physics; he later received a doctorate based on his published research. After his graduation, he taught at Columbia University. In the 1930s, he worked for Shell Oil in Houston. When he was forced to retire, he started working for the U.S. Geological Survey.

In 1956, Hubbert predicted in a public paper that the amount of domestic oil produced would soon go through a maximum. To make this prediction, he used mathematics like that given in Sect. 1.3. His prediction was too pessimistic because new reserves were discovered, including those in Alaska and in the North Sea. However, his work made all realize production would eventually reach a maximum. So far, that maximum has not occurred, and we continue to find and use more fossil fuels with technologies that turn difficult extraction into commercial reality. Proven reserves have remained remarkably constant.

Drake, Rockefeller, and Hubbert had little in common. Today, Drake would probably be a venture capitalist. Rockefeller was a consummate businessman. Hubbert oscillated between the university and the oil business, using mathematics to understand the industry. Still, by 1980, the oil business was in sharp decline. Large sections of the industry are now run by businessmen, not by oil men. Esso (Standard Oil of New Jersey) merged with Mobil (Standard Oil of New York) to make ExxonMobil. Everyone focused either on finding oil or on making higher value-added products than gasoline. General Electric sold its chemical business to Saudi Aramco. DuPont left the commodity chemical business to grow seeds. The Dow Chemical Company sold its commodity chemical business to Kuwait, who then paid a billion-dollar penalty to back out of the deal. The future of the business was grim.

And then, along came George Mitchell (1919–2013). Mitchell was a petroleum engineer and owned an independent energy company. He was also a real estate developer, known especially for The Woodlands community in Houston. He was convinced that existing gas wells could produce more natural gas by hydraulic fracturing ("fracking"). In the fracking process, water, detergents, and sand are mixed into a slurry and then jammed into an oil and gas well to fracture the rock, as shown in Fig. 3.4 (European Environment Agency 2021). The fractures let more liquid and gas escape. Around 1997, Mitchell bet the farm—that is, he put everything he owned into this effort—and he almost went bankrupt; but like a western movie, fracking began to produce cheap oil and gas in massive quantities just in time to save Mitchell's business. The price of natural gas came down by 75%. To put this in perspective, imagine

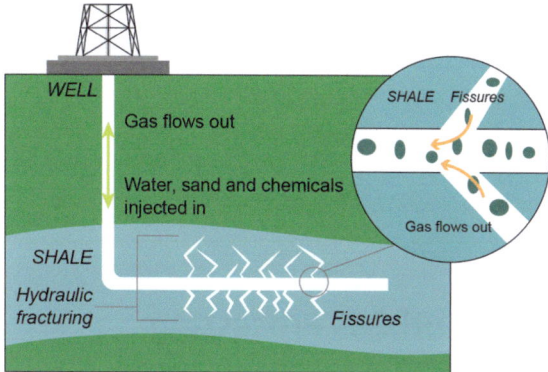

Fig. 3.4 How fracking works. A slurry of soapy water and sand is pumped into an existing well to fracture the rock, releasing gas and oil (European Environment Agency 2021)

that the cost of cars or of steak or of college tuition were to come down 75%. That is what happened in the oil and gas business.

Reflection: What is fracking in physical terms and in chemical terms? Why is it important? Is it more environmentally abusive than strip mining? (See Zuckerman 2013).

In 1997, we asked an oil company president how long this sudden bonanza would last. He smiled broadly and said, "a very long time". When we asked how long a "very long time" would be, he said ten years. Indeed, nearly 30 years after this conversation, the fracking bonanza continues, driven by even more technical advances, both in exploration and in production. Because of these advances, the U.S. has been the largest oil producer in the world since 2015, now producing 50% more oil than even Saudi Arabia or Russia.

Reflection: Though fracking is common in the U.S., it is controversial and not widely practiced internationally, where many countries continue to rely on coal. Because of this, the U.S. increasingly exports its coal to other parts of the world. Globally, coal usage has remained largely flat over the past 15 years. Is this situation good?

3.4 Oil and Gas Processing

Imagine we have crude oil, a black liquid whose viscosity is 10–500 times that of water. The liquid contains molecules with five or more carbon atoms, because the molecules with one to four carbon atoms are gaseous and are in natural gas. We want to convert this crude material into products that can be sold. Some of these products are shown in Fig. 3.5 (Data from U.S. Energy Information Agency 2024). As the figure shows, over 80% of these products are fuels: gasoline, fuel oil, and jet fuel. Less than 20% are chemicals, which range from lubricants to asphalt. However, we should stress that these smaller amounts of chemical products can sometimes be worth more per mass than the fuels.

3.4.1 Two Key Processing Steps

To convert crude oil into products for which there is a market, we first distill the crude oil into different fractions. The "lighter" fractions, which contain 5–12 carbon atoms per molecule, boil at low temperatures and can go directly to gasoline. The "heavier" fractions have higher boiling points and require chemical reactions to convert them to more valuable forms.

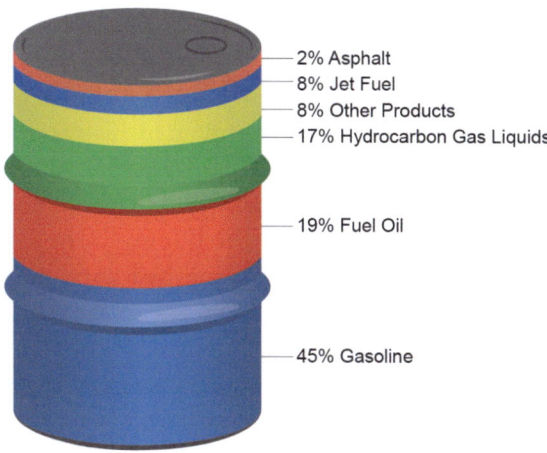

Fig. 3.5 Products made from crude oil. These are dominated by fuels, but the greatest value can be in chemicals (Data from U.S. Energy Information Administration 2024)

Distillation. Distillation is the most important step in the purification of crude oil. In its simplest form, distillation is based on a column—a large pipe—mounted vertically and filled with inert material, like stones. Liquid is boiled at the bottom of column ("the still"), vaporizing the more volatile components. These more volatile components flow up the column and are condensed by heat loss into the cooler surroundings. When the vapors reach the top of the column, they are condensed by external cooling with water. Part of this condensate is put back into the top of the column, and the rest is collected as the product. The more volatile "light ends" come out the top of the column and are a starting point for gasoline; the less volatile "heavy ends" exit at the bottom of the column and are a source of asphalt.

Distillation has been an important process for over two thousand years. Its original form is often credited to Maria the Jewess, who lived around 100 CE in Alexandria, Egypt.

Little is known about Maria. She is credited as the inventor of the double boiler, remembered in French as a *bain marie,* that is, as "Marie's bath." Marie's work was extended by Jabir (722–815 CE), who spent most of his life in Kufa, in modern-day Iraq. Jabir, who may have been Christian, Arab, or Persian, is often credited with the invention of the alembic still. In the lobby of a hotel in Cadiz, in southern Spain, we once saw a carefully polished antique copper alembic used as a garden planter. No one in the hotel knew what it had been.

Distillation uses about half of the energy consumed by the chemical industry, so improving the efficiency of distillation has been a major research goal. Because the efficiency is only about 20% of the minimum required, this improvement has been an enormous research effort. This effort produced some results: in the 1830s, the French recognized that the efficiency could be dramatically improved by pouring part of the condensate back into the top of the column. But additional efforts over the last 200 years have not been as successful.

Perhaps higher efficiencies are simply not possible. In 1823, Sadi Carnot showed that anything involving a hot and cold region will inherently be inefficient because heat will only flow from hot to cold. In 1995, Rakesh Agrawal, a professor at Purdue University, recognized that an additional inefficiency occurred because one component of a mixture would only diffuse from high concentration to low concentration. Those in this area now suspect the efficiency of distillation may already be close to its theoretical limit. We will explore this limit in more detail in Chap. 4; for the moment, we assume that major reductions in energy use by distillation are unlikely.

Chemical Reactions. Once we have separated the crude by distillation, we next need to increase the value of the "heavies," those compounds which cannot be easily distilled. These compounds often consist of molecules of 20–50 carbon atoms. These larger molecules must be broken into smaller fragments in a chemical reactor, and then can be distilled.

The easier type, the tubular reactor, is a piece of pipe filled with a catalyst. Catalysts are often inorganic materials, like silica or zeolites, which have been impregnated with small amounts of noble metals like platinum. Catalysts do not change what is made or the energy needed to make it, but do change how fast any products are made; catalysts are not changed by the reaction but do change the reaction's speed. However, for petroleum, tubular reactors are only one type.

A more common form of chemical reactor in the petroleum industry is the so-called "fluidized bed" or "boiling bed." In this case, the hot heavy oils are slurried with moving catalyst particles, which together look like the sand on a beach where the waves are just breaking. The particles of sand correspond to the catalyst; the hot petroleum vapors correspond to the seawater. Such a reactor breaks up the large molecules in the feed into much smaller molecules, perhaps containing 2–8 carbon atoms. If we are making fuels, these molecules are too small for gasoline. In that case, the output of the reactor is fed into another reactor to partially restore the size of the molecules. In the net result, the crude oil has been converted from high molecular weight species into lower molecular weight ones, like those in gasoline.

These fluidized bed reactors have different names. Those that make the molecules smaller are called "crackers;" when they also have catalysts, they're called "cat crackers," which has nothing to do with dismembering felines. Cat crackers give off heat; that is, they are exothermic. The reverse case, when the product molecules are larger than the feed molecules, is less poetically called a "reformer" and must absorb heat to run at a significant rate. In other words, reforming reactions are endothermic. Both crackers and reformers are common.

3.4.2 What Processed Oil Is Used for

After separation by distillation and molecular size adjustment by chemical reaction, the remaining liquids are converted into fuels—especially gasoline—and chemical products—including ethylene (CH_2CH_2). If we're using carbon-containing molecules as fuel, as synthetic fibers in cloth, or as a drug, we will be using molecules and energy from fossil fuels.

Fuels. As summarized in Chap. 1, 70% of oil is used for transportation and 24% is used in industry. Only 3% is used in residential and commercial settings, and a negligible amount is used for generating electricity. For transportation, oil-based fuels have replaced coal, the key to the Industrial Revolution. Most fuels contain carbon and hydrogen with various ratios of each. When these fuels are burned, they produce primarily CO_2 and water, which may be water as liquid or as steam. The details of this combustion are central to supplying public needs for energy.

Table 3.1 shows the amount of heat released by the combustion of various fuels. The first column gives the fuel itself. The second shows its empirical chemical formula. The last columns show the amount of heat released by complete combustion of the fuel per unit mass and per unit volume under atmospheric conditions. The energy density of the fuel, both in mass and volume terms, is important because it dictates the way we use that energy.

Table 3.1 Heats of combustion of some fuels

Fuel	Formula	Heat of combustion	
		MJ/kg	MJ/m^3
Hydrogen (gas)	H_2	141.8	11.4
Methane (gas)	CH_4	55.5	35.8
Natural Gas (gas)	CH_4–C_3H_6	52.5	35.7
Propane (gas)	C_3H_8	50.4	1000
Octane (liquid)	C_8H_{18}	52.5	36,500
Toluene (liquid)	C_7H_8	48.0	36,800
Kerosene (liquid)	$C_{12}H_{26}$–$C_{15}H_{32}$	46.2	37,900
Diesel (liquid)	$C_{10}H_{20}$–$C_{15}H_{28}$	44.8	36,700
Gasoline (liquid)	C_4H_{10}–$C_{12}H_{24}$	46.5	34,000
Coal (solid)	$CH_{0.3}$–$CH_{0.8}$	32.0	38,000

Except for hydrogen and coal, the amount of heat released per unit mass is remarkably independent of the initial size and structure of the fuel molecule. Everything has a heat of combustion of around 50 MJ/kg. Likewise, the amount of energy released per unit volume by combusting liquid fuels or solid coal is also remarkably constant, around 37,000 MJ/m^3. As expected, the amount of energy released per unit volume of gaseous fuel is much lower than that of liquid fuels. Mass energy density, volumetric energy density, and cost affect how that fuel is used. For example, transportation vehicles like planes, trains, and automobiles need to keep their weight low and volumes small. They rely mostly on liquid fuels. Energy users like home water heaters, industrial furnaces, and power plants are stationary and need fuel delivered to them. They use gaseous, liquid, or solid fuel.

When we design an energy system, we will begin by assuming the combustion of any hydrocarbon fuel gives about the same amount of energy per mass of fuel. When we have taught similar material to undergraduates, we typically take a month in which we go over many more details. For this inquiry into energy, we are ignoring many complexities.

This is concerning. The details are certainly important: for example, the heats of combustion would be lower if we based our estimates on making water vapor and not on making liquid water. The combustion will be affected by whether the gases are present in stochiometric amounts or take place in excess oxygen. The results will depend on whether carbon monoxide is formed. The final temperature depends not only on the amount of oxygen but the amount of nitrogen present whenever the fuel is burned. In some cases, we won't have enough oxygen to burn all the fuel. In other cases, we will have more than enough oxygen, but the reaction won't be complete, so some fuel will remain in the product. When we think about our energy system, we will need to be concerned about these details, but for now, we are simply trying to work out an approximation that will let us begin.

Chemicals made using oil. Many chemical products are based on oil, as suggested by Table 3.2. These examples include plastics, detergents, drugs, and textiles. Textiles are important examples: nylon is an imitation of silk; polyacrylics are wool substitutes used in fleeces; and polyesters imitate cotton. The exception not made from oil is rayon, made from wood pulp.

Polyethylene represents about one-third of the total market for plastics. The ubiquitous supermarket bags are made from polyethene; so is agricultural mulch, wire and cable insulation, squeeze bottles, toys, and housewares. Polyethylene costs about 20% more per ton than oil. As oil prices rise, polyethylene prices also rise. Unlike fuels, where alternative energy sources are available, other raw materials for cheaply making the chemicals in Table 3.2 are harder to imagine.

Table 3.2 Chemicals from oil and gas in the U.S.

Chemical	Formula	Amount (in Mt)	Uses
Ethylene	CH_2CH_2	25	Plastics, packaging, and other chemicals
Propylene	C_3H_6	14	Plastics, foam cushions, carpet
Benzene	C_6H_6	7.7	Styrene, nylon polymers
Acetic acid	$C_2H_6O_2$	2.3	Plastics, including polyvinyl acetate
Ethylene oxide	C_2H_4O	3.5	Antifreeze, thickeners
Formaldehyde	CH_2O	4.4	Foam insulation, particle board glue
Methanol	CH_3OH	3.5	Industrial solvent Acrylic acid
Acrylic acid	$CH_2CHCOOH$	1.0	Textile fibers, detergents
Ammonia	NH_3	12	Fertilizer; explosives

There are some important chemicals that are not oil-based. Cement is based on calcium oxide, made from the reaction:

$$CaCO_3(s) \rightarrow CaO\ (s) + CO_2(g)$$

where the (*s*) and (*g*) remind us that these materials are solids or gases. This reaction releases CO_2 in two ways: first from the decomposition of the lime ($CaCO_3$), and second from the heat required to make the reaction run, typically supplied by burning more coal, coke, or natural gas.

In the same way ammonia (NH_3) is made from nitrogen and hydrogen, one of our favorite research topics:

$$N_2 + 3H_2 \rightarrow 2NH_3$$

which does not seem to depend on oil and gas. However, the hydrogen used in this reaction is most often made from natural gas, so making ammonia produces CO_2 indirectly.

3.4.3 Making Hydrogen from Natural Gas

One special reaction is the manufacture of hydrogen, which is central both to the production of chemicals from non-sustainable hydrocarbons and perhaps for the development of new sustainable infrastructure. In other words, hydrogen is basic both to our current non-sustainable society and perhaps to our future sustainable society. We want to outline the manufacture of non-sustainable hydrogen here; we will come back to sustainable hydrogen later.

Hydrogen is currently made from methane, the principal component of natural gas. The process, called "steam methane reforming," is how over 96% of hydrogen is made today. It has two steps:

$$CH_4 + H_2O \rightarrow CO + 3H_2$$

$$CO + H_2O \rightarrow CO_2 + H_2$$

The first reaction produces a mixture of carbon monoxide and hydrogen called "synthesis gas" or "syngas"; the second reaction, called the "water–gas shift reaction", uses the carbon monoxide and water as steam to make more hydrogen. Overall, for every mole of methane, these reactions make four moles of hydrogen and one mole of CO_2. Steam methane reforming is widely practiced industrially and is the cheapest way to make hydrogen today. The CO_2 is currently released into the atmosphere, and thus technologies such as CO_2 capture and storage, discussed later in Chap. 8, are being commercialized. Other approaches, such as electrolysis of water to make hydrogen and oxygen, are also being investigated, though they remain expensive.

Even though hydrogen gas does not have any color, you may hear about different colors of hydrogen. These colors refer to the process used to make the hydrogen—*green* for water electrolysis using renewable energy, *gray* for steam methane reforming, *blue* for steam methane reforming with carbon capture and storage, *black* for hydrogen from coal, *pink* for water electrolysis using electricity from nuclear power, and *white* for naturally occurring geologic hydrogen. There are more colors, but they all refer to different pathways of making hydrogen.

3.5 Hydrogen and Ammonia as Fuels

Hydrogen is sometimes mentioned as a clean fuel because it can react with oxygen to make water and release energy without emitting CO_2. But hydrogen is a gas and thus its volumetric energy density is low, making it challenging to transport from where it's manufactured to where it's used. Likewise, ammonia is also sometimes mentioned as a clean fuel since it can be decomposed back to nitrogen and hydrogen to further react with oxygen to form water. In some sense, ammonia can be thought of as a hydrogen carrier, which is much easier to liquify and therefore has a high volumetric energy density.

This sparks a question. Rather than an economy powered largely by fossil fuels, why not power the economy by hydrogen and potentially ammonia? After all, hydrogen and ammonia react with oxygen to form water and nitrogen, and so these could potentially be some of the cleanest fuels possible. The chief issue facing this approach is where we get the hydrogen. Referring to the colors from the previous section, 96% of hydrogen made today is gray. It's made by steam methane reforming, which results in CO_2 emissions into the atmosphere. One option to mitigate such emissions is to add carbon capture and storage to make blue hydrogen, but that adds costs and still relies on fossil fuels. Green hydrogen, made by electrolysis using renewable energy, would be better because it does not release CO_2. But green hydrogen made today is more expensive than blue hydrogen and even gray hydrogen. As a result, considerable research is underway to improve performance and reduce the cost of making different "colors" of hydrogen and ammonia.

Reflection: White hydrogen could potentially be present in the earth's crust in large quantities, similar to geologic reservoirs of oil and gas. If this turns out to be true, is it worth extracting and using? How could it change society?

3.6 Conclusions

Crude oil and natural gas are formed by the deposition of biological material on the floors of ancient seas. This deposition occurs over millions of years. Crude oil is produced by drilling a hole in the ground and pumping the oil up from the hole. Before this oil can be used, there are two main processing steps: distillation and cracking. Both are well developed. The bulk of oil processed in these ways is used for gasoline, but a substantial fraction (around 10%) is used for chemicals, including polymers and drugs. These chemicals will continue to be made from hydrocarbons.

Like coal, hundreds of years of supply of natural gas and crude oil remain in the earth. Whether or not these are profitably extractable depends on the economics of supply and demand. As demand increases, innovations like fracking have increased supply. Over the last century, society's demand for energy, both as heat and as work, has grown substantially and has largely been met by fossil fuels because of their properties. But using fossil fuels as an energy source and releasing CO_2 into the atmosphere is not sustainable. Over the next several years, we need to develop and deploy a global energy system that does not emit CO_2 into the atmosphere while providing all the energy that society needs.

References

European Environment Agency. 2021. *Shale Gas Extraction Through Hydraulic Fracturing*. Accessed 1 Dec 2025. https://www.eea.europa.eu/en/analysis/maps-and-charts/shale-gas-extraction-through-hydraulic-fracturing

Energy Institute. 2025. *Statistical Review of World Energy*, 74th ed.

U.S. Department of Energy. 2008. Accessed 1 Dec. 2025. https://web.archive.org/web/20080511161548/http://fossil.energy.gov/education/energylessons/oil/oil4.html

U.S. Energy Information Administration. 2025. *Petroleum Supply Monthly*.

Xu, C., and Laura Bell-Hammer. 2024. OGJ survey shows global oil reserves increase while natural gas reserves decline. *Oil and Gas Journal*.

Zuckerman, G. 2013. *Breakthrough: The Accidental Discovery that Revolutionized American Energy*. The Atlantic 6 Nov 2013.

Part II

Part II Towards Where We Need to Be

4

Energy Transformations (The Hard Part of the Book)

Our next task is to set the stage for finding and using sustainable energy. In the earlier chapters, we reviewed the amount of energy we will need to sustain our current society. We explored where that energy currently comes from and how it is being used. We saw just how big a problem we have. In this chapter, we explore scientific limits on what we will need. We will later identify sustainable energy resources that may meet these needs.

Sometimes the limits on energy are dignified as "The Laws of Thermodynamics." The first law is that energy can be neither created nor destroyed, that the total energy we have on hand must be the amount that we put in minus the amount that we take out. The first law of thermodynamics is sometimes paraphrased by saying that "We get what we pay for."

The second law of thermodynamics is another important limit. It says that energy does not flow from a low-energy state to a high-energy state unless we do some work. In casual terms, heat does not flow from a cold region to a hot one without effort. In more scientific terms, defined below, this is because the disorder, i.e., the entropy, of the entire universe is always increasing: overall, things do always get worse. Sometimes, this is paraphrased by saying, "There is no free lunch." There is a third law of thermodynamics, which states that the entropy of a perfect crystal at absolute zero temperature is equal to zero. That is, there is no disorder in a perfect crystal at absolute zero temperature.

We discuss these ideas in more detail in this chapter. They are more difficult, so some readers may wish to skip to later chapters, where the discussion is in more familiar terms. Before you give up, however, consider a simple example to see what is involved.

Imagine we have a hamper of folded laundry. The hamper has folded underwear separated from paired socks. It has carefully folded pajamas. Everything is ready to be put away in the bureau. But then, in clumsiness, we spill the hamper onto the bed so that all the carefully folded clothes are now a jumble. The energy of those clothes hasn't changed much. They were at room temperature to begin with, so tipping them over doesn't change their temperature more than a very small amount. The clothes were at roughly the same level that they were before, so changes in their height haven't altered their potential energy significantly.

However, we must now do some work to put these clothes back in order so that we can put them away. That work must change some sort of energy in the clothes. What has changed is not the clothes' temperature or their height, but some different kind of energy, which includes their organization, that is, their order. This order is characterized by a new idea, entropy.

4 Energy Transformations (The Hard Part of the Book)

To discuss this and related ideas, we first talk about balances of physically observable energies in Sect. 4.1. Section 4.2 gives a brief history of the study of energy. Section 4.3 reminds us that we can efficiently convert mechanical energy into thermal energy, but we cannot as efficiently convert thermal energy into mechanical energy without doing work. This difference is due to a new quantity, entropy. In Sect. 4.4, we explore the meaning of entropy and how it explains the energy needed to sort a pile of jumbled clothes. In Sect. 4.5, we discuss similar energies locked into the chemistry, like that in fuels. After all, even if they are at the same temperature, the energies of crude oil, of explosives, and of sugar must be different. Finally, in Sect. 4.6, we review these ideas by detailing the original heat engine, the Carnot cycle. The harder ideas in this chapter can be mastered, but with effort; the hardest ideas are in italics, which can be skipped if desired.

4.1 Overview of Physical Energy Balances

Our discussion of energy transformations is most often framed as energy balances, which can be tricky. To aid understanding, we will begin with mass balances, which are easier. These say that for any given system,

$$\text{mass left inside} = \text{mass in} - \text{mass out}$$

In other words, mass is conserved. The amount of pasta in the kitchen pantry is equal to the pasta put in minus the amount taken out. We can think about this in terms of a bank account. The amount of money that we have left in our account is equal to the money we put in minus that we have taken out.

Likewise, when it comes to energy for any given system,

energy left inside = energy in − energy out

Because there are many types of energy—kinetic, potential, electrical, thermal, chemical, internal, and others—we need to be careful to include all forms when making energy balances around a system. Heat and work are the modes of energy transferred across the system boundaries.

These balances are more difficult when chemical reactions are involved. For example, the amount of carbon retained by your body is equal to the mass of carbon taken in as sugars, carbohydrates, fats, etc., minus the mass that is excreted as CO_2 or other waste forms. But the amount of sucrose eaten is not equal to the amount of sugar excreted, because some of the sugar will be digested and metabolized; that is, it will undergo a chemical reaction. Such reactions liberate or consume energy, making energy balances more complicated.

In the next few sections, we focus on energy that comes from physical sources that can change forms but do not involve chemical reactions. In the last section, we add chemical reactions to the mix and explore these complexities in more detail.

4.1.1 Electrical Energy

Electrical energy is the simplest, measured in joules or—if we're interested in power—in watts. If we have 10 kW of electrical power, we can use this either in a space heater to produce 10 kW of heat or in a machine to do 10 kW of work. The amount of energy in joules is equal to the number of watts multiplied by the number of seconds the heater or the machine is on.

4.1.2 Potential Energy

Potential and kinetic energy are two more common forms of energy. Often, the potential energy is the mass m times the change in height h times the acceleration due to gravity g:

$$\text{energy change} = mgh = \rho V g h$$

where ρ is the density of our system; V is its volume; g is the acceleration due to gravity, 9.8 m/s^2; and h is its change in height. For example, the energy required to raise 1 m^3 of water by a height of 0.3 m is:

$$\text{energy change} = 10^3 \text{ kg/m}^3 \times 1 \text{ m}^3 \times 9.8 \text{ m/s}^2 \times 0.3 \text{ m}$$
$$= 2.94 \times 10^3 \text{ kg m}^2/\text{s}^2 = 2.94 \text{ kJ}$$

A cubic meter of water dropping 0.3 m releases nearly 3 kJ of energy; in other units, it releases 0.0011 hp hr.

4.1.3 Kinetic Energy

Kinetic energy is the energy of a mass when it's in motion. It is just one-half the mass m times the velocity v squared. For a cubic meter of water accelerated from rest to two meters per second, the energy change is:

$$\text{energy change} = \left[(1/2)mv^2 - 0\right] = (1/2)10^3 \text{kg}(2 \text{ m/s})^2$$
$$= 2 \times 10^3 \text{ kg m}^2/\text{s}^2 = 2 \text{ kJ}$$

The Jamaican runner Usain Bolt, one of the fastest humans to have ever lived, has been recorded at speeds up to 44 km/

h. His 94 kg body must provide at least 7 kJ of energy to achieve such astonishing speed.

4.1.4 Energy Conversion

Converting energy between different forms is important, and its complexity can be demonstrated by a simple example. Imagine you are riding a bicycle along a hilly road. If you and bicycle have a combined mass m of 100 kg and the hills are 10 m high, the potential energy from a hilltop to the hill bottom is:

$$\text{energy change} = mgh = 100\,\text{kg}\left(9.8\,\text{m/s}^2\right)(10\,\text{m})$$
$$= 9800\,\text{kg}\,\text{m}^2/\text{s}^2 = 9.8\,\text{kJ}$$

If we ride the bicycle down the hill, then the potential energy—how high we are—will be converted into kinetic energy $(1/2)\,mv^2$—how fast we go—as follows:

$$(1/2)mv^2 = (1/2)(100\,\text{kg})\,v^2 = 9800\,\text{kg}\,\text{m}^2/\text{s}^2$$
$$v = 14\,\text{m/s}$$

or about 30 mph. This is consistent with our experience. Of course, because there is friction both in the bicycle and due to the surrounding air, the actual speed will be less, but the idea is still the same.

Next, imagine that we are moving at 14 m per second at the bottom of the hill. When we can now ride up the hill without pedaling: the kinetic energy at the bottom of the hill will be converted back into potential energy at the top of the hill. Without much work, we'll wind up at about the same height. Thus, we can convert potential energy into kinetic energy and back again. We could also convert either

4 Energy Transformations (The Hard Part of the Book)

of these forms of energy into electrical energy with little loss.

Instead, imagine that at the bottom of the hill, when we're going at 14 m per second, we slam on the brakes. When the bicycle stops, the brakes and bicycle rims have gained 9800 J of thermal energy. If we feel the brakes and bike rims after we stop, they will be warm. We have converted kinetic energy into thermal energy.

Finally, imagine having applied the brakes to stop the bicycle, we want to go back to the top of the hill again. If we shout, "Energy, convert!" we won't go up the hill. We're going to have to apply more work to the bicycle through the pedals.

We see that we can convert many kinds of energy into thermal energy, but we can't convert thermal energy back into potential or kinetic or electrical energy without additional work. We understand that energy is conserved: that the energy into the system minus the energy out equals any energy left inside. We also understand that mechanical energy can be efficiently converted into thermal energy. We do not understand why thermal energy cannot be as efficiently converted back into mechanical energy. There is something very different happening. Remember George Orwell in *Animal Farm*: "All animals are equal, but some are more equal than others." Here, all energy seems equal, but some forms are more useful than others.

This incomplete understanding is where we were two hundred years ago. Our society was at a crossroads. Then, we understood that the Industrial Revolution had begun and that the key to this revolution was the steam engine. We did not understand why different steam engines had dramatically different efficiencies. Since then, scientists have been studying energy more deeply, eventually developing the science of thermodynamics. Because this understanding took over a century to develop, you shouldn't be disheartened if you don't understand it right away.

4.2 Genesis of Thermodynamics

Finding the additional rules grew out of the history of coal-fired steam engines. These engines, the core of the industrial revolution, began with the work of Newcomen (1712) and matured with the inventions of Watt (1775) (Morris 2012). But even as these engines became common, they took widely different amounts of coal to do the same tasks. The engines had very different but disappointingly low efficiencies. No one knew why.

The situation for energy two hundred years ago was like what existed in digital processing 50 years ago. We understood that computers could handle a large amount of numerical data in the 1970s. We are now surprised by the idea that computing can make artificial intelligence a reality 2020s. We understood that digital technology was good for accounting but not for natural language. We needed a revolution in our thinking, and we made one through computer science.

Two hundred years ago, we had a similar problem in energy: we needed to understand steam engines at an entirely new level. This understanding came primarily from three people: Sadi Carnot, Rudolf Clausius, and Ludwig Boltzmann. The ideas of these three men provided an intellectual basis for the steam engine and carried us from an agrarian society to our industrial society today.

4.2.1 Sadi Carnot (1796–1832)

The reasons efficiencies were low were first delineated by a French army lieutenant, Nicolas Léonard Sadi Carnot. Sadi Carnot was the son of a major political figure, Lazare Carnot, who during the French Revolution rose from a typical middle-class background to become a member of the Committee of Public Safety and a close collaborator of Robespierre during the Reign of Terror. After Robespierre was executed, the senior Carnot was the first major French politician to support Napoleon Bonaparte. He reorganized the French army, mixing green recruits and experienced soldiers to make the army more effective. When Napoleon became emperor, he made Carnot the equivalent of the Minister of Defense. When Napoleon was finally defeated, the senior Carnot was exiled.

The new French government, which was right-wing, now had a problem with the son, Sadi Carnot, a lieutenant in the army. The government felt the younger Carnot could become a rallying point for French radicals, so it assigned him to remote posts in Algeria. In 1819, he left active service and lived on his pension. He left the army in 1828.

Carnot's great book entitled *Reflections on the Motive Power of Fire* was published in 1824, where he ignored what really happened in a steam engine. Instead, he imagined what happened if a fluid—like air—was cycled between hot and cold temperatures. Carnot's cycle is central to the energy transformation of heat to work; it impacts how society uses energy. Understanding the cycle requires careful thought, which we will undertake in the next section. For now, we show one key result from Carnot's work: the maximum efficiency of *any* heat-to-work engine is limited not by design or engineering, but instead by thermodynamics. This maximum efficiency η turns out to be:

$$\eta = (T_{hot} - T_{cold})/T_{hot} = 1 - T_{cold}/T_{hot}$$

where the engine operates between a hot temperature T_{hot} and a cold temperature T_{cold} expressed in absolute temperature units such as kelvin. Kelvin is just another temperature unit defined as Celsius temperature plus 273.15. Temperature differences are the same when measured in °C or K. The kelvin scale is useful because it is an absolute temperature scale, where zero kelvin is defined as the lowest possible temperature of any system, often called absolute zero.

The maximum efficiency η of an engine is defined as what we want—the work energy performed by the engine—divided by what we pay for—the heat energy fed into it.

Any engine will have a maximum efficiency significantly less than 100%. For example, for a steam engine absorbing heat at 200 °C (473 K) and discharging it at 100 °C (373 K), the maximum efficiency will be limited to:

$$\eta = (473 - 373)/473 = 0.21$$

The greatest possible efficiency that this heat engine can have is 21%, considerably less than 100%, with the remaining 79% being discharged to the environment. Practical engines have an even lower efficiency. In Carnot's time, it could be as low as 3%. This is a huge waste, but this waste is thermodynamically required for engines to convert heat to work. We will return to this potential waste again and again, accepting that it is limited by thermodynamics, and cannot be improved by any new research or new technology.

Even though Carnot is now recognized as one of the founders of thermodynamics, the abstract nature of his analysis made this work unappreciated, indeed ignored, until the insights of Clausius, 30 years later. By then, however, Carnot had died from cholera in 1832 at age 36.

4.2.2 Rudolf Clausius (1822–1888)

The next giant in the study of energy was Rudolf Clausius, born in Köslin, Pomerania, then part of Prussia, but is now in Poland. The son of a Protestant clergyman, Clausius studied physics at the University of Halle, formed from the merger of several schools with the University of Wittenberg, where Martin Luther taught and Hamlet studied. After his PhD in 1848, Clausius taught at Berlin, at the ETH in Zurich, and finally at the University of Bonn. He married, fathered six children, and after his first wife died in 1875, married again and fathered a seventh child. He died two years later.

Clausius realized there was something missing from Carnot's arguments. To be sure, Carnot did explain why the efficiency of steam engines was dramatically less than 100%, but he did not forbid flowing of heat from cold to hot or spontaneous unmixing. If we mix hot water and cold water, we know we get warm water, and we accurately can use energy balances to estimate how warm it will be. But we also know that warm water will not spontaneously separate into hot water and cold water. There is nothing in Carnot's analysis that forbids this.

Clausius developed additional rules that supplement the intellectual structure Carnot had developed to explain the low efficiency. In 1865, he argued there must exist a new function defined as a small amount of heat δQ divided by the absolute temperature T. Clausius called the new function the entropy S. He asserted that:

The energy of the universe is constant.
The entropy of the universe tends to a maximum.

This new function led to the definitions of different forms of energy, like Gibbs free energy, which we will use shortly. Still, while Carnot and Clausius explained why steam energy was inefficient, they could not explain why warm water didn't at least occasionally separate spontaneously into hot and cold water. That was left to Boltzmann.

4.2.3 Ludwig Boltzmann (1844–1906)

The third giant in thinking about entropy was Ludwig Boltzmann, a native of the Austrian city of Graz, who spent some of his happiest years there. He was something of a feminist, at least by the standards of his time: he advocated strongly for educating women and married one pupil whose career he aided; he later supervised Lise Meitner, a discoverer of nuclear fission whose denial of a Nobel Prize is one of the greatest sexist scandals of that prize's history. In the 1880s, Boltzmann provided the statistical basis for entropy, as "statistical mechanics," which is the standard taught today. Boltzmann pointed out that he could not absolutely forbid unmixing into hot and cold; but he could show that it was statistically improbable that this could ever occur. Working on the basis of James Clerk Maxwell's kinetic theory of gases, Boltzmann showed that the entropy S is simply related to the partition function Ω, the number of different arrangements in which the molecules could arrange. This assertion, accepted today, was summarized by the equation:

$$S = k_B \ln \Omega$$

where k_B is a constant, called Boltzmann's constant. This equation is Boltzmann's tombstone. He died a suicide, depressed by the lack of acceptance of this result, the gospel of today.

The work of these three men—Carnot, Clausius, and Boltzmann—gives us the foundations of the science we now call thermodynamics, including how heat energy transforms to work energy. This next section is challenging, and thus it's in italics that you can skip if so desired. However, it can be understood with minimal mathematics and is foundational to understanding why the conversion of heat energy to work energy is thermodynamically limited.

4.3 Converting Heat to Work

The enormous history of the study of energy arises from the Industrial Revolution, that is, on turning heat energy into work energy. In more specific terms, the history centers on burning coal to make steam and in using the steam to run steam engines that convert heat to work.

We want to make these historical ideas accessible to those who have not studied thermodynamics in depth while summarizing the enormous intellectual tradition in this area. In this, we are remembering how the Galapagos Islands informed Darwin's ideas of evolution; we want you readers to be informed about thermodynamics. Over the last three centuries, many engineers have developed engines that operate in a cyclic manner but have less than 100% efficiency. Coal-fired electric power plants are typically 30–40% efficient. Gasoline-powered cars are about the same. Your own body is about 25% efficient: only about a quarter of the calories you eat are turned into the energy of your moving bicycle or using your brain. The rest of the energy is discarded to the environment as heat, including that of sweat. We now show how Carnot and Clausius developed these ideas.

4.3.1 The Carnot Cycle

To begin, imagine a gas contained in a piston like that shown in Fig. 4.1. We assume the gas in the piston follows the ideal gas law:

$$PV = nRT$$

where p is the pressure of the gas, V is its volume, n is the number of moles of the gas, R is the gas constant, and T is the

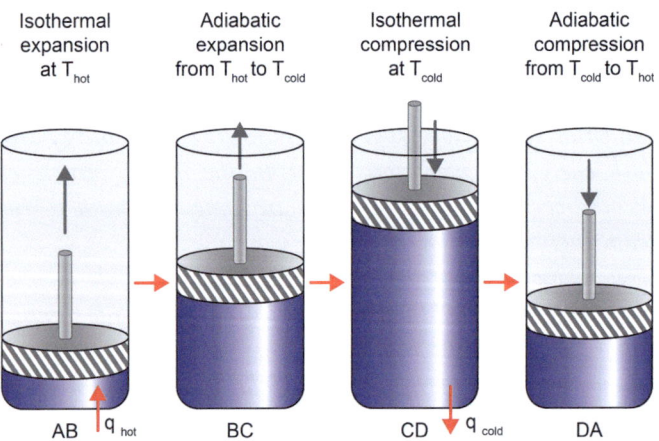

Fig. 4.1 The Carnot cycle. This frictionless cyclic engine, of two isothermal steps and two adiabatic steps, has an efficiency much less than one hundred percent

4 Energy Transformations (The Hard Part of the Book)

absolute temperature. Because the piston is sealed, the number of moles n of gas stays constant.

We also assume that in this ideal engine, there is no friction between any moving parts; everything is made from thin and light materials that conduct heat perfectly and instantaneously; the system can be insulated completely when desired; the gas is always homogeneous; and the system has no inertia. We assume that heating or cooling the gas can be accomplished by contacting the outside of the cylinder with a large thermal reservoir that is always at T_{hot} or T_{cold}. We also assume that by covering the cylinder with insulation, the gas does not gain or lose heat.

We do not care how fast heat is transferring, how fast the piston is moving, or the overall speed of the engine. We care only that heat can be transferred and that the piston can move, even if its movement is infinitely slow. A process that undergoes a change in slow, small steps is called reversible; such a process is the most efficient in converting heat to work, as we shall soon see. Reversible processes can be extremely slow, but thermodynamics is not about speed; it's about energy.

Now we use this heat engine in the four-step Carnot cycle shown in Fig. 4.1 and track the condition of the gas in two ways, as shown in Fig. 4.2. The left diagram shows the pressure p and volume V of the gas at any point in the cycle suggested by Carnot. The right diagram shows the temperature T and the entropy S of the gas as suggested by Clausius. We should stress that the diagrams shown in Fig. 4.2 are not simple. As students, each of us remembers the confusion and the mathematical complexity these diagrams caused us. If this seems opaque, you haven't paid your dues of working through the Carnot cycle in the way that Carnot originally suggested.

The cycle starts at point A, where the temperature of the gas in the piston is infinitesimally less than the temperature of a hot thermal reservoir at temperature T_{hot}. This infinitesimally small temperature difference means that heat will flow

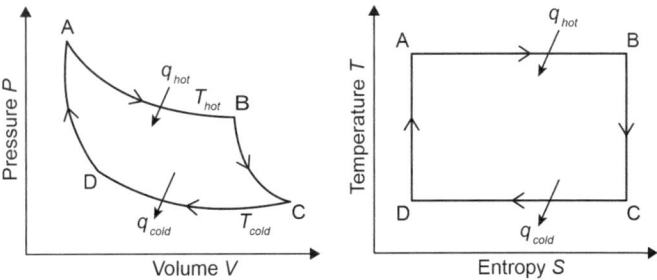

Fig. 4.2 Two equivalent ways to show a Carnot cycle. The pressure P—volume V plot on the left is due to Carnot; the temperature T—entropy S plot on the right is due to Clausius

reversibly from the hot reservoir to the gas in the piston. The gas gains heat q_{hot} and expands its volume until point B such that the temperature of the gas is always kept at a constant T_{hot}. In other words, step AB is a reversible isothermal expansion of the gas at temperature T_{hot}. Because the expansion is isothermal, the ideal gas law states that the product PV is also constant, and thus the pressure P must decrease during the expansion. Thus, in Fig. 4.2, the curve AB is a hyperbola on the PV diagram on the left and a flat horizontal line at constant temperature on the TS diagram on the right.

In the second step from B to C, we remove the heat reservoir and insulate the cylinder. But now, we incrementally reduce the pressure outside the cylinder. For example, if there were a bag of sand on top of the piston head, the pressure could be reduced by removing one grain of sand at a time. As the pressure drops, the piston moves reversibly to expand the volume of the gas until it reaches point C. But this time, the gas cools as it expands, and neither P, V, nor T is constant. This type of frictionless expansion, where no heat leaves or enters the system, is called "adiabatic" in which PV^{γ} turns out to be constant. This new term γ is the ratio of heat capacity at constant pressure to that at constant volume. For an ideal gas,

4 Energy Transformations (The Hard Part of the Book)

the difference between the constant pressure heat capacity and constant volume heat capacity is equal to the gas constant R. For diatomic gases such as oxygen and nitrogen, γ is about equal to 7/5.

Clausius noted that, if an adiabatic process is also reversible, then it must also be isentropic—the entropy must be constant. We allow the reversible adiabatic expansion in step BC to continue until the temperature of the gas inside the piston is infinitesimally higher than the temperature of a cold thermal reservoir at temperature T_{cold}. Thus, in Fig. 4.2, the curve BC is at constant PV^{γ} in the PV diagram on the left and a straight vertical line at constant entropy in the TS diagram on the right.

But now the engine is stuck. To keep the engine going, we need to bring the piston back to its original state. But how? The only way is to contact it with a cold thermal reservoir at T_{cold}. Because we stopped the adiabatic expansion at point C when the gas temperature was infinitesimally more than T_{cold}, heat can still flow reversibly from the cold gas to the cold reservoir. The gas loses heat q_{cold} at a constant temperature T_{cold} until it reaches point D. Step CD is an isothermal compression at T_{cold}. Importantly, because the gas is cooler in step CD than in step AB, the gas molecules have lower velocity and thus lower kinetic energy. As a result, compression in step CD will be easier and take less energy than in step AB. q_{cold} will be smaller than q_{hot}. Like curve AB in Fig. 4.2, curve CD is a hyperbola on the PV diagram on the left and a flat horizontal line at constant temperature T_{cold} on the TS diagram on the right.

At point D, we stop removing heat from the gas and insulate the cylinder. We then reversibly compress the gas by increasing the pressure outside the cylinder. This could be done by repeatedly adding one grain of sand at a time, reversing the process in step BC, and going back to the top of the piston head. This is an adiabatic compression, during which the gas heats up until *it reaches a temperature incrementally lower than the temperature T_{hot}, the exact conditions we started with in step A. That is, in step DA, the gas is compressed adiabatically from T_{cold} to T_{hot}. Like curve BC in Fig. 4.2, curve DA is at constant PV^γ in the PV diagram on the left and a straight vertical line at constant entropy on the TS diagram on the right.*

Because the gas started and finished at the same temperature and pressure over the four-step cycle, Carnot realized its energy was unchanged. The gas had no memory: it only knew it was in the same conditions where it had begun. But the heat q_{hot} added at T_{hot} and q_{cold} rejected at T_{cold} are not equal. The difference between the two is the work produced by the piston:

$$W = q_{\text{hot}} - q_{\text{cold}}$$

Note that heat must be rejected from this cycle, meaning q_{cold} must be greater than zero. This means that the work W must be less than the heat q_{hot} provided to the cycle. In other words, the engine's efficiency in converting heat to work must be less than 100%. But Carnot took it one step further. He showed that because the Carnot cycle is reversible, that is, frictionless, it has the highest possible efficiency of any heat-to-work engine, which we derive next.

> *Reflection: In the example above, sand was removed at a higher elevation and added back at a lower elevation. Discuss practical ways to convert that potential energy to other forms of energy.*
> *Reflection: Explain the Carnot cycle to a friend who has no scientific education. (Most people have trouble doing this.)*

4.3.2 Engine Efficiency

We can pictorially represent any heat-to-work engine like that shown in Fig. 4.3. The engine absorbs thermal energy from a hot reservoir, discharges some of it to a cold reservoir, and converts the difference between the two into work.

An engine's efficiency η is the work that it produces divided by the heat that we pay for:

$$\eta = \text{work/heat} = [q_{in} - q_{out}]/q_{in} = 1 - (q_{out}/q_{in})$$

For an ideal gas operating in a Carnot cycle, this is equivalent to:

$$\eta = (T_{hot} - T_{cold})/T_{hot} = 1 - (T_{cold}/T_{hot})$$

This efficiency is much less than 100%, even though the heat engine has no friction. In Carnot's time, it could be as

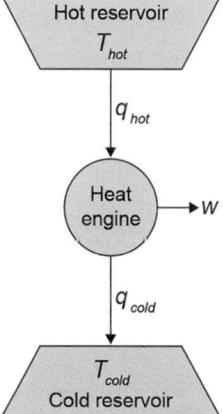

Fig. 4.3 A heat engine. A heat engine converts thermal energy into work

low as 3%. Carnot showed that using a higher temperature heat source and a cooler temperature heat sink will increase the efficiency of an engine operating in a cycle. The most practical heat sink is the ambient environment. As a result, engines are often designed to operate as high a temperature as possible, limited only by the materials of their construction. Today, some of the highest efficiency engines at large scales are power plants, where turbines are made from expensive high-temperature metal alloys. Such power plants can have efficiencies of upwards of 60%. Though much higher than in Carnot's time, they are still limited by Carnot efficiency at much less than 100%.

Remember that we are dealing with The Laws of Thermodynamics. We often refer to "first law" and "second law" efficiency in this context. A steam turbine can have a near 100% first law efficiency when converting the thermal energy in steam entering the turbine to the kinetic energy of rotating the turbine blades. But if the cool steam exiting the turbine is heated and returned to the turbine inlet to operate as a cycle, the thermal energy required to reheat the steam will be more than the turbine's work output. This is often called the cycle efficiency or the second law efficiency and is thermodynamically limited.

4.3.3 Heat Pumps

We can also run an engine to convert mechanical work into heating or cooling. Such an engine has a different kind of efficiency, called a "coefficient of performance (COP)," again defined as what we want divided by what we pay. For a refrigerator, where we want cooling, this is

$$\text{COP} = T_{\text{cold}}/(T_{\text{hot}} - T_{\text{cold}})$$

4 Energy Transformations (The Hard Part of the Book) 91

For a heat pump, where we want heating, we get

$$\text{COP} = T_{\text{hot}}/(T_{\text{hot}} - T_{\text{cold}})$$

Unlike the normal efficiency, the COP can be greater than one because these machines use work to move heat between hot and cold reservoirs. Still, the ideas are similar to those for efficiency: both are what we want divided by what we pay.

4.4 Entropy and Energy

Thus far in this chapter, we have described how different kinds of mechanical energy can be interconverted. We note that mechanical energy can be efficiently changed into thermal energy, but thermal energy cannot efficiently be changed into mechanical energy. This inefficiency of thermal energy conversion into mechanical energy is illustrated by observing that heat does not spontaneously flow from cold to hot or, equivalently that warm water does not spontaneously unmix into hot water and cold water. We have explored the reasons for this, uncovered by Carnot, Clausius, and Boltzmann. This has been some of the most difficult material in this book, so if you think it is hard, you are right.

In this section, we expand this observation by exploring entropy and the second law of thermodynamics. We start by sorting clean laundry, when different articles of clothing must be separated. Conceptually, this is equivalent to separating warm water into hot water and cold water.

4.4.1 Entropy as Disorder

To put these arguments on a more physically understandable basis, we again consider a laundry basket of carefully folded clothes—socks, underwear, shirts, and so forth. If we stumble and drop the laundry so the clothes become jumbled, the normal forms of energy would not be changed. The potential energy and the kinetic energy of the laundry are the same. The internal energy is unchanged: the laundry doesn't get much hotter by dropping it. But some important property of the laundry has changed because of the accidental jumbling—it is going to take work to put the laundry into its organized form. This property is entropy, a measure of the disorder of a system.

To simplify the problem, we limit the laundry basket to just shirts and underwear. The change in entropy ΔS can be determined by counting the number of ways shirts and underwear can be arranged on the bed. The higher the number of possible arrangements (which thus leads to higher disorder), the higher the entropy. For a binary system of just shirts and underwear, the change in entropy is given by:

$$\Delta S / nR = -[x_1 (\ln x_1) + (1 - x_1) \ln(1 - x_1)] > 0$$

where n is the total number of articles of clothing, R is a constant (the gas constant), x_1 is the fraction of the clothing which are shirts, and $(1 - x_1)$ is the fraction of underwear. For mixing different species, the entropy change is always positive. For example, imagine we have 20 shirts and 14 underwear. The fraction of shirts x_1 is:

$$x_1 = 20/(20 + 14) = 0.59$$

That of underwear is:

$$x_2 = 14/(20 + 14) = 0.41$$

4 Energy Transformations (The Hard Part of the Book) 93

The entropy change for the 34 articles of clothing is positive:

$$\Delta S/R = -34[0.59(\ln 0.59) + 0.41(\ln 0.41)] = 22.8$$

Mixing increases the entropy, always.

In this case, because the number of pairs of socks and underwear is so small, these equations are not exact. But in molecular systems, the number of molecules is huge. For example, for a tablespoon of water that contains over 10^{23} water molecules, this result is accurate. In the same sense, if the molecules are of different sizes or shapes, or if they slightly react, the results will be different. Still, for our purposes, remember that mixing increases entropy. And because the second law of thermodynamics states that the entropy of the universe is increasing, this also means that mixing is a natural tendency of systems. To reverse this process and unmix the system, we must do work.

4.4.2 Gibbs Free Energy

In 1863, Josiah Willard Gibbs received the first Ph.D. in engineering from an American university, Yale. A few years later, he proposed a function we now call the Gibbs free energy, G, which included terms for both heat and entropy:

$$G = U + PV - TS = H - TS$$

where U is the internal energy, H is the enthalpy (heat), P is the pressure, V is the volume, T is the temperature, and S is the entropy. Gibbs showed that the change in Gibbs free energy between an initial state G_1 and a final state G_2, $\Delta G = G_2 - G_1$, is the amount of available energy to do work. He further noted three interesting limits:

  $\Delta G < 0$: the change will be spontaneous from state 1 to state 2

ΔG > 0: the change will be spontaneous in the opposite direction, from state 2 to state 1

ΔG = 0: there will be no change between the two states; the system is at equilibrium

These limits exist because the Gibbs free energy combines both enthalpy and entropy. In nature, a system will tend to move to a lower energy state. When you hold a ball above the ground and let it go, it will drop because the ball has lower energy on the ground. A protein will reorient itself to the lowest energy configuration. But the universe also tends to higher entropy, as stated by the second law of thermodynamics. Whether a system will spontaneously change or not depends on both energy and entropy, as defined by the Gibbs free energy, with the sign of *ΔG* indicating the direction of the change.

Consider the example of reacting gaseous hydrogen and oxygen (state 1) to make liquid water (state 2):

$$2\,H_2(g) + O_2(g) \rightarrow 2\,H_2O\,(l)$$

The change in enthalpy *ΔH* between the two states is the heat absorbed or released by the reaction. From calorimetric experiments, *ΔH* is measured to be -285.83 kJ/mol, where the negative sign indicates energy is released. Though this negative sign indicates that the system will be at a lower energy state and the reaction wants to proceed, we must also consider the change in entropy. Note that we should expect the entropy change *ΔS* to be negative, because the reaction is making liquid molecules from gas molecules. Molecules in a liquid are more confined and thus more ordered than molecules in a gas that have more space to move around. Additionally, the reaction produces fewer molecules than it started with—two molecules starting from three—which also increases order. Values for entropy of many molecules are tabulated in the literature. The change in entropy can be *calculated from such tables as −163.5 J/K-mol at 25 °C. The negative sign indicates entropy will decrease as we had expected, indicating that the reaction does not want to proceed. The Gibbs free energy change combines both enthalpic and entropic changes to give us a final result:*

$$\Delta G = \Delta H - T\Delta S - 285.3 \text{ kJ/mol} - (25 + 273 \text{ K})$$
$$(-0.1635 \text{ kJ/K-mol}) = -237.1 \text{ kJ/mol}$$

Because ΔG is negative, this reaction will proceed spontaneously at 25 °C—kaboom!

> *Reflection: Explain the following forms of*
> *energy in non−technical terms:*
> *Internal energy U*
> *Enthalpy $H = U + PV$*
> *Entropy S*
> *Gibbs free energy $G = H − TS$*
> *(We challenge every audience with this reflection,*
> *and we often get good answers.)*

4.4.3 Energy of Unmixing

In the previous section, we showed that the Gibbs free energy centers around both energy and entropy. For a system that is mixed that we want to unmix or sort, such as the laundry of shirts and socks, we must do work. But the work we do in unmixing does not result in a change in the enthalpy (heat) of the laundry. Instead, it goes towards decreasing the entropy of the system. The energy of unmixing, which is positive, is exactly equal to the negative of the energy of mixing, which is negative.

This energy of unmixing is important because it is work that we must do to mitigate pollution. If we want to separate carbon dioxide from ambient air, we must do work to do so. The amount of work will depend on the concentration of carbon dioxide. The more dilute the CO_2, the more work we will have to do for each molecule we remove. Just like needles in a haystack, fewer needles in the haystack require more work to find each needle.

To illustrate this further, we imagine that we mix carbon dioxide and air at temperature T; we will pretend "air" is pure, making the good approximation that oxygen and nitrogen are the only components. In this case, no heat is added to or removed from the gas, and thus the Gibbs free energy decrease on mixing $\Delta G(\mathrm{mix})$, is given by:

$$\Delta G(\mathrm{mix}) = \Delta H - T\Delta S = 0 - \Delta S = -T\Delta S$$
$$= -T\,[-nR][x_1 \ln x_1 + (1-x_1)\ln(1-x_1)] < 0$$
$$\Delta G(\mathrm{mix})/nRT = x_1 \ln x_1 + (1-x_1)\ln(1-x_1)$$

where n is the total number of moles of gas; R is the gas constant; T is the absolute temperature; and x_1 is the mole fraction of species "1" in the gas. The factor nRT has the dimensions of energy, which means that the left- and right-hand sides of the equation are dimensionless, valid for any units used.

The energy change of unmixing is the minimum work W required to separate the gases. This is just the negative of the energy change on mixing:

$$W/nRT = \Delta G(\mathrm{unmixing})/nRT = -\Delta G(\mathrm{mixing})/nRT > 0$$

This minimum work does not depend on the technology which we choose; no matter how much research we do, no matter how smart we are, we will not do less work. This minimum energy has practical consequences for estimating the cost of reducing

4 Energy Transformations (The Hard Part of the Book)

the CO_2 in the atmosphere to reduce global warming, or the cost of recovering lithium from the sea to build better batteries.

To illustrate, imagine we take one mole of oxygen weighing 32 g and four moles of nitrogen, each weighing 28 g. We mix these at a temperature of 25 °C to make 144 g of synthetic air containing 20 mol% oxygen and 80 mol% nitrogen, where the "mol%" reminds us that the stated composition is moles, not grams. Then the minimum energy needed to separate this mixture into the pure gases is:

$$\Delta G(\text{unmix})/nRT = -x_1 \ln x_1 - (1-x_1) \ln(1-x_1) > 0$$
$$\Delta G(\text{unmix})/[(5 \text{ moles})(8.31 \text{ J/mol K})(25 \text{ K} + 273 \text{ K}]$$
$$= -0.2 \ln(0.2) - (1-0.2) \ln(1-0.2)$$
$$\Delta G(\text{unmix}) = 6.2 \text{ kJ}$$

This energy is the minimum any process would take to carry out this separation of oxygen and nitrogen in air. We cannot do better. Just as the Carnot efficiency limits the engine, the energies involved in mixing constrain our operations: concentrated gas solutions always mix to make dilute solutions. We must do work to return to where we began.

We can carry these ideas further by thinking about distillation. Distillation involves both temperature and concentration differences. It is limited not only by heat flowing from hot to cold but also by solutes diffusing from concentrated to dilute. We guess that the efficiency of distillation will almost certainly be reduced by the need to supply free energy to effect the separation. Distillation in the oil business consumes a million barrels of oil per day. Agrawal argues that distillation is about 10% efficient, so this seems a wonderful opportunity for reducing energy use. However, the constraints reported here mean, unfortunately, that distillation is already close to its minimum energy.

Reflection: Salt is present in the sea at about 35 g/l. Assuming most of the salt is sodium chloride, what is the mole fraction of ions from salt in the sea? What is the minimum energy for taking the salt out of the sea?

Reflection: How does the second law of thermodynamics prohibit all perpetual motion machines?

4.5 Chemical Energy Balances

We now turn from energy stored physically (perhaps as high temperature or high purity) to energy stored chemically, that is, within chemical compounds. We will largely be concerned with energy stored in either a fuel or electrochemical energy in a battery. In both cases, we will want to convert this chemical energy into some other form, most often mechanical or electrical energy. At present, however, we are largely concerned with how much energy is available.

4.5.1 Combustion

When we burn any fuel, such as hydrogen or coal, we must carefully specify the conditions of the combustion. We must choose a temperature and pressure for the fuel. We must decide which chemical reactions we expect to occur. We need to specify whether we expect the combustion to produce steam or liquid water. Finally, we must choose a basis for the combustion: do we want to know how much energy we produce *per mole* or *per mass* or *per volume*?

4 Energy Transformations (The Hard Part of the Book)

These issues are illustrated by the simplest possible fuel, hydrogen. We can burn hydrogen in oxygen, producing water vapor:

$$H_2 + 0.5\ O_2 \to H_2O\ (g)$$

Of course, the temperature produced by this combustion will depend on where we start: on whether we burn the hydrogen in pure oxygen or in air; and whether we burn it in excess air or in the stochiometric amount (that is, the exact amount needed by the chemical reaction). These specifications are frequently important. However, in the context of this book, we are largely interested in relative values, so the details of this combustion are deferred. While they certainly are essential to any detailed energy analysis, they are less important for the approximate values here.

To illustrate the ideas involved, we will burn the fuel at 25 °C in a stoichiometric amount of oxygen to make only water vapor and gaseous CO_2. In the case of pure hydrogen, we will obtain an energy of around 286 kJ/mol, or in units frequently used in North America, of 61,500 BTU/lb. The values for hydrogen and for some other common fuels are listed in Table 4.1.

Several complexities in this table require additional explanation. First, the energy shown is the enthalpy change

Table 4.1 Heats of combustion for some fuels

Fuel	ΔH, kJ/mol	ΔH, BTU/lb
H_2	286	61,500
C	394	14,100
CH_4 (methane)	802	22,000
C_8H_{18} (octane)	5074	19,100
Coal (anthracite)	–	14,000

The values in this table assume water is formed as a vapor

ΔH, which is still another energy that includes any pressure and volume changes produced by the reaction. The enthalpy turns out to be a more convenient and reliable measure of energy in the description of combustion, and its use has no major effect in most of the cases we consider here. Second, the units are given in the table both in kJ/mol and in BTU/lb. The conversion between these units is straightforward using the factors given in Sect. 4.4.2. For example, for hydrogen in kJ/mol, we have

$$(286 \times 10^3 \,\text{J/mol})(\text{mol}/2\,\text{g})(454\,\text{g/lb})(\text{BTU}/1055\,\text{J}) = 61{,}500\,\text{BTU/lb}$$

which is the other value shown in the table.

This table implies that whenever water is made, it is produced as vapor, not as liquid. The ΔH values shown are therefore called Lower Heating Value (LHV). If water were produced as a liquid, the ΔH values shown would be larger, called the Higher Heating Value (HHV). We have assumed that any carbon combustion produces only CO_2 and not carbon monoxide (which is frequently produced in combustion); if we allowed carbon monoxide, the ΔH values given would be smaller. Finally, whenever possible, we have given the values as ΔH per mole, not per mass. This is not possible for anthracite coal, which is not chemically well-defined but often given the empirical formula $CH_{0.5}$.

Energy released by chemical processes like combustion is normally larger than the energy produced by physical changes. Of course, we prefer to use compounds as fuel that release a lot of energy. Still, we can now anticipate a future problem: replacing these fuels with more sustainable resources is going to be hindered by the lower energy density of many sustainable alternatives. Sustainable fuels may not have energy densities as high as those obtained with hydrogen or fossil fuels.

4.5.2 Electrochemical Energy

Energy can also be collected and stored chemically in a battery, so the electrical energy will be available when needed. Such a battery has four parts: a cathode, an anode, a separator, and an electrolyte. During discharge, the cathode consumes the electrons produced in the cell. The anode supplies those electrons. The electrolyte lets ions pass from anode to cathode so electrons can pass through an external circuit. The separator prevents chemicals in the cathode and anode from reacting directly without a current. Such batteries are a promising method for storing energy generated periodically by solar and wind.

One easy comparison of different battery chemistries is their energy density, as shown in Table 4.2. Two features of this table are striking. First, lithium-based batteries have a higher energy density than the traditional lead-acid battery. This partly reflects the difference between the molecular weight of lithium (7) and that of lead (207); a lithium atom is nearly 30 times lighter than a lead atom. This difference is one reason for the push towards lithium-based batteries for electric cars, which we will discuss in detail in Chap. 8. Other battery chemistries are also higher than the lead-acid standard. We also note that all these electrochemical options have dramatically lower energy densities than regular old gasoline. It is not an accident that our current energy grid is based on gasoline, and not electrochemistry.

Table 4.2 Energy densities of common batteries

Type	Energy density, kJ/kg
Lithium-ion	940
Lead acid	400
Nickel metal hydride	500
Vanadium	50
Gasoline (for comparison)	45,000

4.6 Conclusions

In analyzing processes, we will often make balances on mass and energy. The mass balances are straightforward, though the effect of chemical reactions can be a complication. Energy balances are more difficult, partly because energy has different forms which can be converted from one form to another. For example, the kinetic energy of a pitched baseball is transferred into heat when it strikes a catcher's mitt. However, energy balances are sometimes stymied by real but complicated limits. In the case of the catcher's mitt, turning the heat back into the moving ball requires some work. The catcher can't command the ball to spontaneously leap back to the pitcher; the heat can't be turned into kinetic energy.

Carnot explained these energy limits by using a four-step cycle that converts heat into work at the highest possible efficiencies. Later, Clausius and Boltzmann used the concept of entropy to explain why heat only flows from hot to cold—unless we do some work to reverse the process. For CO_2 in flue gas, CO_2 always moves from high to low concentration—unless we do some work. These observations in turn imply that the entropy of the universe is increasing—the second law of thermodynamics—setting limits on how we develop energy for our society.

Reference

Morris, C.R. 2012. *Dawn of Innovation: The First American Industrial Revolution.* Public Affairs. New York: Perseus Books Group.

5

Is Sustainable Energy Feasible? Steady Energy Resources

We begin exploring if sustainable decarbonized energy is even possible: can we collect enough energy from renewable sources to support our current way of life? In trying to answer this question, we assume that human energy, animal energy, and natural fire are negligible relative to our total demands. In other words, we assume that humans picking up bricks, oxen pulling plows, and fires warming a house do not supply a large fraction of the energy needed for our current way of life. We are trying to not only meet the energy demands of today's society, but also those of the future. Over the next few chapters, we shall investigate how these demands can be met in a decarbonized fashion, which is the energy transition.

Our conclusion will be that the future energy system will likely need all types of energy resources, with sustainable energy being an important and growing component of a broader energy system. This newly sourced sustainable energy can be less expensive than some conventional energy sources. However, if society chooses to adopt this sustainable path, the sources of energy we use in the future will certainly be a different mix than what we use at present. The future energy may not all be in the forms we expect. Before we discuss these details, we need to reconsider how we compare energy from different sources.

5.1 Why Energy Conservation Is Hard to Understand

Discussing fossil fuel energy and sustainable energy is hard because not all energy is created equal. In earlier chapters, we summarized energy from coal, oil, and natural gas. We reviewed energy units and converting between them. In Chap. 4, we introduced the issue of converting heat energy from fossil fuels into work energy, like motion or electricity. This conversion is inefficient because heat flows from hot to cold. It only flows from cold to hot if we do extra work. This inconvenient truth is summarized by entropy, a hard idea that explains the limits of energy conversion.

Furthermore, different industries use different units to measure energy—joules, kilowatt hours, British Thermal Unit (BTU), calories, and many more. Power, which is energy per unit time, is also expressed in different units—watts, horsepower, and BTU/hr. Sometimes units also have subscripts to distinguish heat from work. A power plant may combust natural gas to generate 1000 MJ of thermal energy, but if its heat-to-work conversion efficiency is around 35%, then it will generate only 350 MJ_e of electrical energy, with the remaining 650 MJ_t thermal energy discharged into the surrounding environment. That same 350 MJ_e of electrical energy can be expressed as 97.2 kWh.

5   Is Sustainable Energy Feasible? Steady Energy …

This complexity makes everyone—governments, engineers, private citizens—have trouble talking about this subject. These troubles begin with estimates of energy needed, like the one that we gave in Fig. 1.1. Such charts include solar and wind, which produce electricity directly, but they also include biomass and geothermal energy, which are thermal. Thus, we are including solar-sourced work energy, which is highly efficient, with machines that convert heat to work, which are inherently inefficient, limited by entropy.

Professionals in this area respect these differences but handle them in different ways. Before September 2023, the U.S. Energy Information Administration (EIA) converted the joules of solar and wind energy to an equivalent amount of thermal energy (U.S. Energy Information Administration 2023a). In other words, the EIA treated solar energy as displacing fossil-fuel-fired power plants. They divided the actual energy produced via solar and wind each year by the then current average efficiency of the U.S. fossil-fired power plant fleet, which is about 0.35, to calculate the renewable energy. Starting September 2023, the US EIA stopped doing so and followed other agencies, such as the International Energy Agency (IEA). Most energy organizations now simply add the joules of electrical energy from solar and wind to the joules of thermal energy from fossil fuel sources. Since solar and wind are now a relatively small fraction of U.S. and global energy consumption, making a different conversion has a small effect on estimates of the total energy consumption, at least for now. This may not be true in the future.

Now that we have discussed some of the complexities, consider Fig. 5.1, which is a modification of Fig. 1.1 (Data from U.S. Energy Information Administration 2024). It treats the electric power sector differently for two reasons. First, the electric power sector consumes a large fraction of the energy sources in the U.S, 34.6 EJ, which is about one-third of the total 99.4 EJ of U.S. energy sources. Of that, about 41% is transmitted and used for various end uses, while 59% is lost to the environment in converting heat to work and transmitting electricity over long distances.

The second reason why the electric power sector is treated differently is that electricity is not a primary energy source like oil, natural gas, coal, renewables, or nuclear. Rather, electricity is made from such energy sources. Electricity is an energy carrier, a means to transmit energy from the location where it's created to a location where it's used. Because electricity moves through wires at nearly the speed of light, it enables energy to be transferred practically instantaneously over very long distances. These attributes have made electricity the backbone of our modern society.

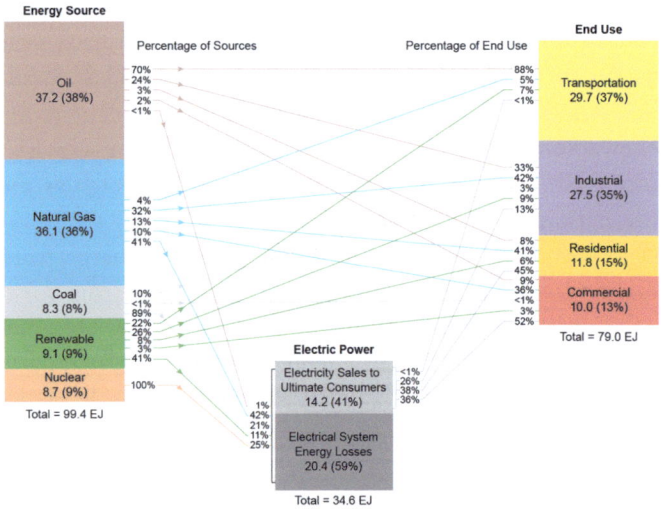

Fig. 5.1 Energy use, including some issues of heat energy and work energy. U.S. Energy consumption totaled 99.4 EJ in 2024, or about 10 kW per person (Data from U.S. Energy Information Administration 2024)

But electricity also has a challenge: it cannot be stored. If we want to use electricity later, it must first be converted to another form of stored energy, such as thermal, chemical, or mechanical, which can then be converted back to electricity.

Sustainable energy, such as solar and wind, generates electricity directly, and thus, sustainable energy must be consumed as soon as it is generated. But our society uses energy even if the sun isn't shining or the wind isn't blowing. If our new energy system relies heavily on sustainable energy, it must also include energy storage, which will add cost. We will explore many of these nuances throughout the remainder of this book, and ultimately look for least-cost pathways that can provide a decarbonized future. The answer is not always clear, even to energy experts, which is why this conversation is hard.

5.2 Sustainable Energy Is Cheap

We now explore how much sustainable energy will cost. After all, while we plan to change society's source of energy, we do not want to change dramatically the lifestyle we currently enjoy. We hope to avoid going back to a dark, cold, medieval world. When we try to estimate these costs, we find sharply differing values, calculated sensibly but with different technical and financial assumptions, such as the amount of time in a year the unit would be operated, efficiency of the unit, fuel costs, interest rates, financing terms, expected life of the unit, and so on. We've tried to choose values that are a consensus of these varying estimates. We sometimes will give the estimates in kilowatt hours, not joules, because these more common units allow quick comparison of many sources, including those that convert heat to work.

The cost of energy includes the costs of both capital and operation and maintenance, as detailed in Table 5.1 (U.S. Energy Information Administration 2023b). Operation and maintenance (O&M) includes variable costs such as fuel and machine maintenance that can fluctuate with plant operation, and fixed O&M costs such as rent and insurance, which are constant regardless of plant operation. These data show that onshore wind and photovoltaic solar have some of the lowest capital and fixed O&M costs, with zero variable O&M costs. Consequently, these are now the first choice for the lowest cost electricity. However, these are also intermittent sources, producing electricity only when the wind blows and the sun shines. In contrast, natural gas plants are also low capital but have higher variable and fixed O&M costs. They can be dispatched on demand, but also emit CO_2. Adding carbon capture and storage reduces CO_2 emissions, but at increased cost. These differences mean any detailed answer to our question "Which form of energy is best?" will not be simple.

The capital and operating costs can be combined into a term called the levelized cost of energy or electricity (LCOE), where the capital cost is amortized over a fixed time and added to the operating and maintenance costs. It is conceptually like amortizing a home loan over a fixed time and adding it to the periodic cost of maintaining the home. The levelized cost of owning your home includes monthly loan payments consisting of principal and interest, as well as other costs such as utilities, property taxes, insurance, and periodic house repair. If you are going to buy a house, you need to consider at least the levelized cost, not just the mortgage payments.

LCOE can be very low for some technologies, but that gives only one part of the picture, because that technology may not meet all energy demand. Our energy demand varies across days and seasons, across different industries,

Table 5.1 Capital and operating costs for different ways to make electricity

Technology	Capital cost, $/kW	Operating and maintenance		Levelized cost of electricity, $/MWh
		Variable, $/MWh	Fixed, $/kWy	
Dispatchable				
Coal	$4507	$5	$46	$89
Coal with CCS	$7319	$12	$67	
Natural gas combined cycle	$1176	$2	$14	$43
Natural gas combined cycle with CCS	$3140	$7	$31	
Biomass	$4998	$5	$142	$77
Nuclear, small modular reactor	$8349	$3	$107	$71
Geothermal	$3403	$1	$154	$37
Fuel cell	$7291	$1	$35	
Resource-constrained				
Wind, onshore	$2098	$0	$30	$31
Wind, offshore	$6672	$0	$124	$100
Solar, photovoltaic	$1448	$0	$17	$23
Conventional hydropower	$3421	$2	$47	$57
Capacity resource				
Natural gas combustion turbine	$867	$5	$46	$129
Battery storage	$1270	$0	$46	$117

Costs are for new large plants (U.S. Energy Information Administration 2023b)

and across different locations. We need to consider the total cost of matching all this varied energy demand with an equally matched decarbonized energy supply. And we must do so not just for today's energy landscape, but for the next few decades. This is not an easy exercise and involves many assumptions, which we will discuss further in Chap. 9.

Before 2010, energy from solar panels cost more than twice that from natural gas: the price of natural gas had dropped both because of deregulation and because of fracking. But in the last decade, the cost of electricity from solar panels and from wind has become the cheapest. Parenthetically, electricity generated from natural gas, included in this table as "Natural Gas Combustion Turbine," is expensive partly because it is only used periodically to cover shortfalls from the other sources. It is used only to cover peaks of high demand. The increasing power demand for data centers and artificial intelligence is putting upward pressure on the costs for all types of power generation, particularly natural gas combined cycle.

That solar and wind have become so inexpensive does not seem to have been recognized by the public. You may remember Moore's Law, which is a hypothesis expounded in 1965 by Intel's co-founder, Gordon Moore, and modified in 1975. Moore asserted that the number of transistors on an integrated circuit would double approximately every two years. Remarkably, this assertion has been held for decades as the performance of electronics increased, and their cost decreased. A similar statement was asserted by Theodore Wright in 1936. Wright asserted that the cost for airplanes decreased by a constant amount for each doubling of cumulative airplane production. This price pattern could be true for the manufacturing of other items and services, but the pattern will probably be truer for new technologies and less so for more mature alternatives.

Wright's and Moore's Laws are explored by Fig. 5.2 (Roser 2025), which shows the price of electricity versus the cumulative installed capacity for different technologies from 2010 to 2019. This figure reflects not only the price but also the changes in price caused by technical experience and scale. As the figure shows, the price of energy produced from coal drops only 2% when the installed capacity goes from one million megawatts to two million megawatts. This increase in capacity does not teach anyone how to use coal more efficiently. Though not shown in the figure, natural gas power plants behave like coal power plants. Making electricity from fossil fuels is mature. In contrast, the price of solar energy has dropped by over 70% as the technology is used on a larger scale. These changes are the natural maturations of the different technologies.

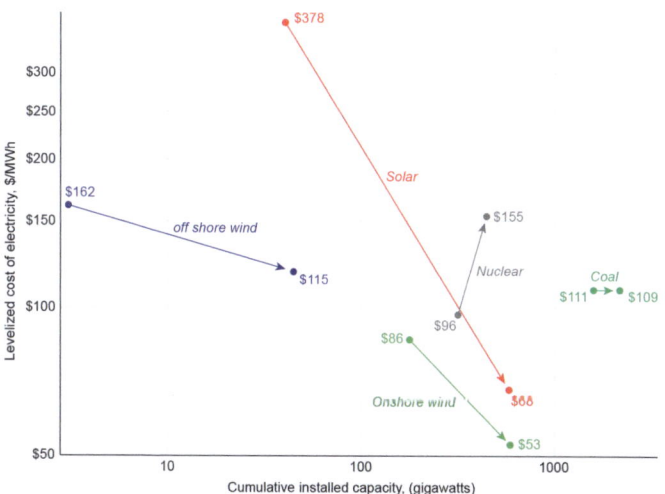

Fig. 5.2 Energy costs versus installed capacity from 2010 to 2019. As coal capacity increases, its cost per megawatt does not drop. As solar capacity increases, its cost drops dramatically (Roser 2025)

Using electric power from the sun and wind involves a different infrastructure. Regions with constant bright sunshine or strong winds tend to be in areas with low population, like deserts and plains. People don't want to live in these hot, windy places. Moreover, the sun doesn't shine all day, and the strength of the wind varies considerably both diurnally and seasonally.

Thus, we should distinguish between steady and periodic energy sources. In the remainder of this chapter, we focus on steady sources. In Chap. 6, we discuss periodic energy sources like the sun and wind. In Chap. 7, we move to ways in which periodically available energy can be stored and energy variations can be cushioned. For now, we focus on hydroelectric and nuclear power because these are the largest, steadiest sustainable resources.

5.3 Hydropower Can Be Steady

Getting energy from water is centuries old, the basis of the water mills in fairy tales. This old technology takes water at a higher elevation, runs it through a mill to make mechanical work or electricity, and discharges the water at a lower elevation. Obviously, no CO_2 is emitted in the process, though there could be CO_2 released during the manufacturing, maintenance, and repair of the watermill. In many cases, hydroelectric energy is steadily made, though seasonally fluctuating.

The energy ΔE available from the water is the mass moved times the acceleration due to gravity times the drop in height:

$$\Delta E = \rho V g \Delta h$$

where ρ is the density of the water; V is its volume; g is the acceleration due to gravity, close to 9.8 m/s^2, and Δh is the drop in height. Sometimes it's more convenient to calculate the power, or energy per time. For example, if we have a small water mill with a flow of a cubic meter per second dropping two meters, the power produced is:

$$(1000 \text{ kg/m}^3) \, (1 \text{ m}^3/\text{s}) \, (9.8 \text{ m/s}^2) \, (2 \text{ m})$$
$$= 19{,}600 \text{ kg m}^2/\text{s}^3 = 19.6 \text{ kJ/s} = 19.6 \text{ kW}$$

Simple mills like this were the basis of the Industrial Revolution.

Hydroelectricity made from flowing water remains a reliable way to make electricity. For example, we remember one visit to the Bull Shoals dam in Arkansas in 1971. This mill, built after the Second World War, began operating in 1951. When we visited, the engineer in charge gave us a wonderful tour, stressing the robustness of the entire mill, asserting it had especially reliable instrumentation and control. He showed us his state-of-the-art speed governor. We recognized the design as that developed by James Watt in 1788. The technology really was so good that it flourished for centuries.

But hydroelectricity produces only a small fraction of what we need. As we have discussed, the U.S. consumes 10 kW per person for our population of about 330,000,000. The Bull Shoals dam produces 380 MW or 0.001 kW per person, about 0.01% of our need. The dam is not big enough. We could potentially mitigate this shortfall with more and larger hydroelectric dams. However, as detailed in Marc Reisner's book, Cadillac Desert (1993), the U.S. has already dammed every river with the potential for significant energy production. There is more potential for hydroelectric energy generation in other parts of the world, like Latin America and Africa, but the additional potential is still small relative to consumption. More hydroelectric power will help, but it will not solve our energy problems.

Reflection: We are discouraging the search for more hydroelectricity, saying that all promising rivers have already been dammed. But tides near Boston are three meters and those near London are twice that. Could tides supply significant renewable energy for Boston or London?

Reflection: Another potential source of renewable energy is when the fresh Rhine River water is mixed with the salty North Sea. How much of the Netherlands' energy demand could this mixing supply? See Sect. 4.4 for help in getting started. What would be the environmental consequences of such an energy system?

5.4 Nuclear Power Can Be (Nearly) Steady

Another possible source of energy, which can be steady, and which does not emit much CO_2 during its operation, is nuclear power. Like the windmill, nuclear power plants can emit CO_2 during construction and maintenance, and when mining its uranium fuel. Most agree there is abundant uranium to support an expanded nuclear power industry.

Concerns about nuclear safety seem overstated because the chemistry used to make weapons is different than that used to make electricity. Weapons can use either fission or fusion. For a fission weapon, an "atomic bomb", one must separate the common uranium isotope 238, which is not explosive, from the rarer uranium isotope 235, which can be. Uranium 235 must then be purified, which is difficult because of the high entropy of unmixing and the small mass difference between isotopes. To make a fusion weapon, a "hydrogen bomb," a mixture of the hydrogen isotopes deuterium that contains two neutrons (^{2}H) and tritium that contains three neutrons (^{3}H) are compressed so much that the atomic nuclei fuse, making helium that contains four neutrons (^{4}He) and releasing a high-energy neutron:

$$^2H +^3 H \rightarrow\, ^4He + \text{neutron (energy)}$$

The energy released is enormous. This remains a topic for research as means for carbon-free energy production. Recently, scientists at the Lawrence Livermore National Laboratory in California used 192 laser beams to add 2 MJ of energy into a small pellet of hydrogen isotopes to release 3 MJ. Though a long way from commercialization, fusion may one day lead to decarbonized energy. But not soon.

In contrast, the generation of electricity from uranium-235 is relatively simple. We assemble enough uranium in a sufficiently concentrated form so radioactive decay causes it to get hot—hot enough to boil water. More specifically, we make ceramic pellets containing uranium. Each pellet is about one cubic centimeter (10^{-6} m^3) and has the energy of 0.6 m^3 of diesel oil, that is, 160 gallons. The ceramic pellets, packaged into fuel rods, boil water because of radioactive decay; the resulting steam is fed to turbines like those used to make electricity from coal and natural gas. Regrettably, this energy is subject to a (low) Carnot efficiency.

There are about 100 atomic energy reactors in the United States and 450 worldwide, with some power plants having two or more reactors. Each reactor is about 1000 MW. France produces 70% of its power from nuclear energy. The safety of nuclear power benefits from sharing data through the International Atomic Energy Agency in Vienna. Countries operating nuclear plants, including China, share in this facility; their work has received the 2005 Nobel Peace Prize. Nuclear power causes about 500 times fewer deaths per MW than coal-based power. Future power plant designs are claimed to be both safer and cheaper. The plants could also be modular, assembled in a centralized location rather than onsite, which should reduce their cost. The future for nuclear energy seems bright.

Reflection: Concerns with nuclear safety are legitimate and require careful evaluation. Advocates of nuclear power sometimes compare this industry with airlines, that is, with commercial aviation. Airplane deaths were high at the start of commercial flights, but they dropped as planes became more reliable. Some claim the same is true of nuclear power, and that nuclear technology is now much safer than it was. Is this true? Does this suggest turning to nuclear power on a much larger scale?

One problem with nuclear power is the eventual fate of radioactive waste material. This problem, often called NIMBY (Not In My Back Yard), can be addressed by separating, vitrifying, and storing the waste in a central location, or by storing the waste in place. We now explore this a bit more.

The first step in managing radioactive waste is to make the volume of waste smaller. One method of doing this is incineration; others include ion exchange and liquid–liquid extraction. Liquid extraction involving inorganic phosphates has been most widely practiced. Another method is absorption in inorganic materials like sodium titanate. These concentrated solids are then vitrified, that is, turned into glass. The vitrified solid is placed in a long-term, geologically stable repository. All these techniques are believed to be safe and inexpensive. According to the United States Government Accountability Office, the national repository at Yucca Mountain, Nevada, was abandoned not for technical reasons, but for political and public opposition (GAO 2011).

An alternative method avoids concentrating the waste but mixes it with cement and then guards the resulting concrete block. At the Savannah River Laboratory in South Carolina, this means having a guard for 300 years. Nonetheless, this method is safer and cheaper than concentrating and transporting the waste to a central site. In both cases, the cost of treating the waste is a minor fraction of the value of the electricity made.

Reflection: Finland is heavily committed to nuclear power, especially as a route to replace coal. They have had little public protest, that is, no NIMBY, apparently because they have included all stakeholders from the initial planning stage. Are they wrong? Would Finland's plans work in other countries? (El-Showk 2022).

5.5 Conclusions

Sustainable energy is achievable and can be cheaper than energy based on fossil fuels, especially when considering social costs. Sustainable energy should also always be available; some sources of sustainable energy are not. Hydroelectric power is always available, but there's not much room for expansion. It is also subject to seasonal fluctuations. While hydroelectric energy is attractive for countries in Scandinavia and Africa, it is essentially fully developed in the United States. Nuclear power is stable: designs and modular construction may affect significant cost reductions. Nuclear power remains bothered by problems with waste disposal, leading to political and public opposition.

Hydroelectric power and nuclear energy are two low-carbon options to supply power for our current society. Achieving a society based on sustainable low-carbon energy will also require ways to produce and store electricity from the sun and the wind. These are the subjects of the next chapter.

References

El-Showk, Sedeer. 2022. Final resting place. *Science* 375: 806–810.

Government Accountability Office. 2011. *Commercial Nuclear Waste: Effects of Termination of the Yucca Mountain Repository Program and Lessons Learned*. GAO-11-229.

Reisner, M. 1993. *Cadillac Desert: The American West and Its Disappearing Water*. New York: Penguin Books.

Roser, M. 2025. *Why Did Renewables Become so Cheap so Fast? Our World In Data*. Accessed 1 Dec. 2025. https://ourworldindata.org/cheap-renewables-growth

U.S. Energy Information Administration. 2023a. *Monthly Energy Review, Appendix E: Alternative Measures for the Energy Content of Noncombustible Renewables*. Accessed 1 Dec. 2025. https://www.eia.gov/totalenergy/data/monthly/pdf/MER_E.pdf

U.S. Energy Information Administration. 2023b. *Cost and Performance Characteristics of New Generating Technologies, Annual Energy Outlook*. Accessed 29 Sept 2024. https://www.eia.gov/outlooks/aeo/assumptions/pdf/elec_cost_perf.pdf.

U.S. Energy Information Administration. 2025. *U.S. Energy Consumption by Source and Sector 2024*. Accessed 1 Dec. 2025. https://www.eia.gov/totalenergy/data/monthly/pdf/flow/total_energy_spaghettichart_2024.pdf.

6

Is Sustainable Energy Feasible? Periodic Energy Resources

The early parts of this book emphasized existing sources of energy, especially fossil fuels. While these energy sources have effectively met the world's needs for the last 300 years, they are now held responsible for much air pollution and climate change. For many past years, fossil fuels represented the cheapest forms of energy. While nuclear energy was believed in the 1950s to be the salvation of the future, reservations about nuclear safety and waste have been a significant deterrent, holding the fraction of total energy that is nuclear to below 10%.

Developments in solar- and wind-generated power offer real alternatives to fossil fuels, and they can be exploited for the foreseeable future. These energy sources are now cheaper than traditional fuels. The pace of these developments has astonished everyone. Thirty years ago, the president of a large electrical utility said that no solar cell had ever produced more energy than was required to make it in the first place. This is no longer true: the possibility of building a new society including large amounts of sustainable energy now seems feasible.

The difficulty is that solar and wind power are periodic. We don't want electricity for lights during the day when the sun is shining. We want lights at night. We don't want electrical appliances to shut off when the wind dies down. We want them to be always available. We need to design our new sustainable society to store solar- and wind-generated energy, but before we decide how we're going to store energy, we need to review how this sustainable electricity is made. That is the subject of this chapter.

6.1 Solar Is Ready Now

Solar energy can be used either to make electricity directly or to convert the sun's heat into fuels. Making electricity directly depends on solar cells, and hence on microelectronics. Collecting the sun's heat directly is older, simpler, and less impactful (at least for the moment). However, both deserve consideration.

6.1.1 Silicon Solar Cells

To discuss how electricity is made from sunlight, we first need to discuss how electricity is conducted in metals, salts, and semiconductors. In metals, an electron is not always associated with an individual metal atom. For example, an electron in copper is not associated with a specific atomic nucleus of a single copper atom. Sure, most of the electrons are held strongly in specific orbitals, in the valence band. But the outer electrons in metals are not associated with a single nucleus, but in a conduction band. Electrons in this band range freely throughout the metal. Sometimes the electrons in the conduction band are described as an electron gas, as if the electrons were a separate phase superimposed on the well-organized metal crystals. When an electrical potential is put across the metal, the electrons move and induce a current.

Salts are completely different because each electron is now associated with a specific atomic nucleus. For example, in solid sodium chloride, sodium ions and chloride ions are packed into a rigid (face-centered) cubic lattice. Electrons in the valence bands of the sodium or chloride ions remain fixed. The outermost shell of the chloride ion contains one more electron than there are positive charges in the nucleus, so it has a net charge of − 1. The outermost shell of the sodium ion has one less electron than the number of protons in the nucleus, so that it has a net positive charge of + 1. When an electrical potential is applied across the salt, the electrons do not move: the solid salt is not a conductor.

When the salt is molten or dissolved in water, the melt or the solution conducts a little electricity, but as ions, not as electrons. When a potential is applied across the sodium chloride melt, the positive sodium ions move towards the cathode, and the negative chloride ions move towards the anode. In the same way, in an aqueous solution, individual electrons don't move, but the hydrated ions do. In these cases, the electric current—the rate of charge transfer—is about ten million times slower than that in metals but much greater than in solid salt crystals.

But what happens in a semiconductor like silicon? On the one hand, we understand that an electric current in a metal is due to electron motion. On the other hand, we also understand that the current in a solid ionic crystal is nearly zero. What happens in a semiconductor?

In a semiconductor without light, the electrons are immobile, trapped in their valence bands. The conductivity is small. But when light shines on the conductor, some of the photons can be absorbed, increasing some electron energies enough to raise those electrons into the conduction bands, as shown in Fig. 6.1. The threshold shown in the figure may seem surprising, but similar behavior is frequent. For example, imagine how toothpaste flows when you squeeze the tube. At first, nothing happens, but after a threshold squeeze is reached, flow is easy. Conduction of an electrical current in a semiconductor is similar: if the photons in the light exceed a threshold energy, the electrons can conduct, behaving more like the electron gas in a metal.

To use this effect to make a silicon solar cell, we first make an interface between two regions. On one side of the interface, we add traces of elements with extra electrons for an n-type (negative) doped material; examples of these elements are arsenic and phosphorus. On the other side, we add traces of elements deficient in electrons, for a p-type (positive) material. Examples of these elements are boron and gallium. Such a p–n junction produces a bias in electrical conductance. Now we complete an electrical circuit outside the solar cell, so when light strikes the interface, current will flow through the circuit. We have used light

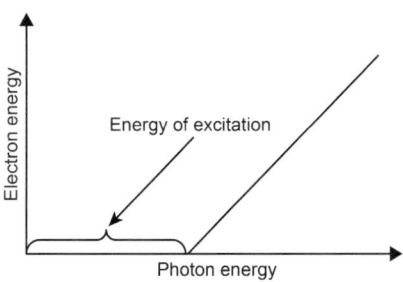

Fig. 6.1 Electron energy versus photon (light) energy. After a threshold is reached, current can flow

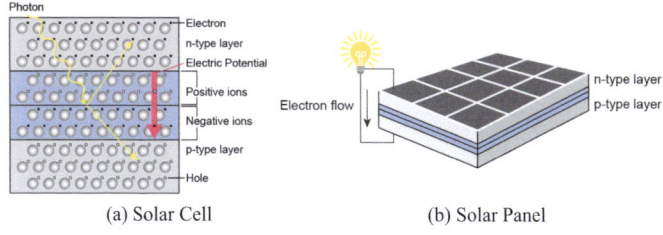

(a) Solar Cell (b) Solar Panel

Fig. 6.2 A Schematic picture of a silicon-based solar cell. a Light which reaches the interface across which a potential is applied produces a current. **b** Thus, light generates electricity (ACS ChemMatters 2014)

to produce an electric current; we have the basics of a solar cell, as shown in Fig. 6.2 (ACS ChemMatters 2014).

It is astonishing that these devices have been developed so rapidly. In the 1960s, solar cells were perhaps 2% efficient; now, they asre over 20% on the commercial scale and over 45% in research labs. Then the cost was over $300 per watt; at a large scale, it is now $1 per watt, 300 times cheaper. Even flexible organic solar cells have shown these improvements: in 2000, their efficiency was just 3%; now the efficiency is near 20%. Solar cells are one key to sustainable solar power. However, the power is generated only when the sun is shining. Energy storage is a huge challenge, as will be detailed in Chap. 7.

Reflection: Sunlight has an incident radiation of about 1000 W/m^2. Solar panels in a city should produce 10% of that, or about 100 W/m^2. If we supply all our electrical needs with solar, we will need a significant land area. Discuss how this could be achieved for a town of 50,000 people and for a city of 2,000,000.

Reflection: Solar power in Minneapolis, Minnesota, yields around 4.9 kWh/m^2/day averaged over the year, but this can be as low as 3 kWh/m^2/day during a cold December *winter and 6.3 kWh/m^2/day during a balmy July summer. On the other hand, solar power in Palo Alto, California, yields around 5.9 kWh/m^2/day averaged over the year, with a low of 4 kWh/m^2/day during a typical December winter and 7.2 kWh/m^2/day during a typical August summer. How will these values differ for a city like London, which is further north?*

6.1.2 Solar Heating for Domestic and Commercial Use

While electricity generated with solar energy is a high-tech undertaking, domestic hot water generated from solar energy is a low-tech, ancient technology. The descendants of early efforts are rooftop water heaters on homes, which typically save individual consumers around $400 per year. Some commercial units use computer-controlled mirrors to focus the sun's energy on a small target. With sufficient mirrors, the target area can get hot enough to boil water to make steam, which is fed to a conventional steam turbine like those used to make electricity from coal. Most elegantly, focused solar energy can melt a salt mixture at high temperatures to make a slurry, a compact way of storing energy. When needed, this melt can be allowed to fuse, releasing energy to make steam to feed a turbine. However, while these technologies represent alternatives to fossil fuels, none is growing rapidly. The capital investment is too great for the power produced, at least at the present time.

6.1.3 Solar Energy for Chemicals

More exotically, solar energy can be focused with mirrors to generate such high temperatures that compounds decompose into their elements directly. For example, water decomposes into a mixture of hydrogen and oxygen. The temperature for significant decomposition is around 3000 K. For comparison, welding uses 3400 K, and the sun's surface is at 5800 K. The elevated temperatures needed are difficult to handle since mixtures of hydrogen and oxygen are explosive. While the mixture would be less dangerous if the two gases were separated, separation at this elevated temperature is difficult. Though this technology is appealing, it is currently underdeveloped and remains speculative. It seems unlikely to become commercially important soon.

Other approaches to making chemicals from solar cells are indirect and have higher commercial appeal. They use solar cells to first generate electricity, which in turn supplies the energy needed to drive chemical reactions, including the dissociation of water into hydrogen and oxygen. Such reactions are endothermic, that is, they consume energy, but when the reactions are reversed, they are exothermic, that is, they release energy. Such an electrochemical system can be engineered to be an energy storage device, a means to store solar energy and release it later. If sustainable energy is going to be our future, such energy storage ideas play an increasingly important role, and we will discuss them in more detail in the next chapter.

6.2 Wind Is Ready Now

Wind has been a source of power since ancient times. The Greeks pumped water using wind energy before 70 CE, and the Persians developed this technology more completely by 600 CE. Their early windmills used canvas sails. In modern times, Charles Brush, an inventor who lived in Cleveland, Ohio, used wind energy to generate electricity: in 1877, his was the first house in Cleveland with electric lights.

To generate electricity from wind, we want large wind turbines as high up as possible to avoid disruption of wind speed caused by surfaces, and as big as possible to collect as much energy as we can. At present, this means the vanes on the turbines are 20–80 m long, made of fiberglass composites. The price for such turbines is a dollar per watt, the same cost as the solar cells, making equal power.

Reflection: Wind generated electricity is said to be much cheaper now than 20 years ago because of better materials. Which materials? Why are these better?

Interestingly, the maximum fraction of power that can be extracted from windmills is about 16/27, a value called the "Betz limit" after one of its inventors. We can derive this result as an example of the approximate engineering analysis basic to many of the conclusions in this book. This analysis is outlined in the next paragraphs. If the somewhat higher level of math is worrying, you can skip this material and just start reading again when the next section starts.

Our approach is to use the conservation of energy: kinetic energy from the wind is transferred to the turbine. In other words, the wind velocity is v upstream the turbine is higher than wind velocity u downstream the turbine, with an average wind velocity (v+u)/2 through the turbine. The power of the wind P_w upstream of turbine is the mass flowrate of the wind multiplied by the kinetic energy of the air molecules per unit mass:

$$P_w = (\rho A v)\,(v^2/2) = \rho A v^3/2$$

6 Is Sustainable Energy Feasible? Periodic Energy ...

where the air density ρ has dimensions of mass per volume ml^{-3}, the swept area A has dimensions of length squared l^2, and the wind velocity v has dimensions of length per time l/t. Thus, the power of the wind has dimensions of energy per time ml^2/t^3, as it should.

The difference between the wind's kinetic energy upstream and downstream the turbine is the power captured by the turbine blades, P_b:

$$P_b = (\rho A(v+u)/2)(v^2/2 - u^2/2) = \rho A(v+u)^2(v-u)/4$$

The velocity difference $(v - u)$ means if the wind upstream and downstream has the same velocity, then v and u would be equal, and no energy would be transferred.

The efficiency η of the wind turbine is the power captured by the blades P_b divided by the power available in the wind P_w.

$$\eta = P_b/P_w = \rho A(v+u)^2(v-u)/(2\rho A v^3) = (1+\zeta)^2(1-\zeta)/2$$

where ζ is u/v, the ratio of downstream wind velocity to the upstream wind velocity. We seek the maximum efficiency, so we assume many values of ζ and calculate the efficiency η. We can tabulate this efficiency as a function of the velocity ratio ζ. Some of the values are:

ζ	η
0	0.5000
0.1	0.5455
0.15	0.5261
0.3	0.5915
0.333	0.5926
0.34	0.5925
0.5	0.5625

The efficiency is greatest when the ratio of downstream wind velocity to upstream wind velocity is one-third. Thus, the maximum efficiency is:

$$\eta_{\text{maximum}} = (1 + 1/3)^2 (1 - 1/3)/2 = 16/27 = 0.593$$

This is the result which we set out to derive. This performance is consistent with actual operations.

An analysis like this seems straightforward but tricky. We can feel we understand every step, but do not really know what is happening. A deeper understanding is easy with calculus. If you understand calculus, you can find these results without the tedium of making tables or graphs. If you're interested, we urge you to do this, but understanding the result does not require calculus.

6.3 Conclusions

At this point, we see how making sustainable energy from solar cells or wind turbines is achieved. While making sustainable electricity from other sources like tides or geothermal wells is also possible, these methods are less developed than sun and wind, which are ready to go now. Making sustainable energy for society does not need to depend heavily on new methods that need dramatic future development. These methods are among the cheapest to manufacture, and do not have much cost to operate. But they produce energy intermittently, when the sun shines or when the wind blows, and thus we need methods for *storing* sustainable energy. Storing solar energy made during the day for use at night or recovering wind energy harvested in the fall for use in the winter does need development. Our immediate problem in making a society based on increasing sustainable energy is less energy harvesting than energy storage, the subject detailed in the next chapter.

Reference

A Solar Future. 2014. *How a Solar Cell Works*. ACS ChemMatters.

7

Sustainable Energy Requires Energy Storage

We know how much energy we are using and how we can decarbonize it. But we must supply this energy not at a constant rate but on demand. This demand varies with the time of day and the time of year. If we want energy from sustainable sources to play a larger role in the energy transition, we must make it available when it's needed and not just when the sun is shining or the wind is blowing.

To focus our discussion, we will consider only energy storage which does not involve the emission of carbon dioxide. This means we will not consider any fossil fuel combustion or manufacturing in this chapter. For simplicity, we also will not consider any CO_2 footprint associated with the energy storage apparatus. We will assume that we are using the sun and wind to make sustainable electricity. Since electricity cannot be stored directly, we want to convert electricity into energy that can be stored mechanically, thermally, chemically, electrochemically, or some other means so we can release it on demand.

7.1 Mechanical Energy Storage

We begin by discussing how energy can be stored in water reservoirs and compressed gas tanks, two examples out of many. The most familiar form of energy storage is a water tower, a standard feature of Midwestern towns with modest changes in elevation. We can see one such tower from the room where we are writing this chapter. It's about 50 m tall, and it has a volume of around 3000 m^3. This tower ensures a steady flow in the town's water supply, so toilets run smoothly when residents come home from their jobs in the evening.

7.1.1 Storing Energy with Water

The water tower stores an amount of energy ΔE:

$$\Delta E = \rho V g h$$

where ρ is the density of the water, V is its volume, g is the acceleration due to gravity, and h is the change in elevation, that is, the "head." This local reservoir contains an energy:

$$\Delta E = \left(1000 \text{ kg/m}^3\right)\left(3000 \text{ m}^3\right)\left(9.8 \text{ m/s}^2\right)(50 \text{ m})$$
$$= 1.5 \times 10^9 \, J = 1500 \text{ MJ}$$

This seems like a lot. But if the 50,000 people in this town use energy at the normal rate of 10 kJ/s, this water tower will only supply energy for 1.5×10^9 J/[50,000 people $(10 \times 10^3$ J/s person)] or 3 s. We will need much more energy storage than this water tower can supply. Conventional water towers like this are an excellent choice for damping variations in domestic demand, but are much too

7 Sustainable Energy Requires Energy Storage

small to be a big part of our society's sustainable energy infrastructure.

Water towers would be more useful for energy storage if they were much bigger. Not surprisingly, this technology has been thoroughly explored. Four examples are shown in Table 7.1; the water tower, which we just discussed, is the first entry in the table. The second entry is related to the Gordon Butte wind farm near Martindale, Montana, on top of a mesa above the surrounding plains. It has six 1.5 MW turbines, each 80 m below the top of the mesa. Because the local wind velocity averages nine meters per second (20 mph), this farm can generate about nine megawatts. This is the maximum generation capacity, called nameplate capacity, which assumes that wind speed is constant without any downtime. Because the wind blows intermittently and all machines have periodic outages, the actual generation of the wind farm is on average about 30% of the nameplate capacity, called the capacity factor. Thus, the energy available is:

$$(6)(0.30)1.5 \times 10^3 \, \text{kW}/(10 \, \text{kW/person}) = 300 \, \text{people}$$

This is only a few people. In Montana, during the times that the wind is not blowing, the electricity is supplied by the grid.

The owners of the Gordon Butte facility would also like to use their location for energy storage. They are building two reservoirs, each of which will hold approximately five million cubic meters of water. One reservoir is on top of the mesa, 300 m above the second reservoir; the second one, at the bottom of the mesa, stores the water. These reservoirs can store:

$$1000 \, \text{kg/m}^3 \left(5 \times 10^6 \, \text{m}^3\right) 9.8 \, \text{m/s}^2 (300 \, \text{m}) = 15 \times 10^{12} \, \text{J} = 15 \, \text{TJ}$$

Table 7.1 Hydroelectric energy storage

Location	Purpose	Power source	Head, m	Reservoir, m^3	Backup
Edina, MN	Water supply	Grid	50	3 k	Grid
Gordon Butte, MT	Electrical surge	Wind	300	5 M	Grid
Bath Power, VA	Load level	Grid	380	34 M	Grid
El Hierro, ES	Supply	Sun, wind	1500	0.5 M	Diesel

This is enough energy for 300 people for:

$$15 \times 10^{12} \, \text{J} / \left(\left(10^4 \, \text{J/s}\right) (300 \, \text{persons}) (3600 \, \text{s/h} \times 24 \, \text{h/day}) \right)$$
$$= 60 \, \text{days}$$

Like the local water tower, this larger water reservoir can damp out oscillations in power generation and demand, but if it were the only energy source, it could only support 300 people for 60 days. Still, we are going to need considerably more storage for larger towns, though this facility can handle a winter cold wave in Martindale, MT.

A larger operation is the Bath Regional Energy Storage Facility near Warm Springs, VA. From the data in the table above, we can show that this system can store around 130 TJ. This huge amount should greatly help ensure a steady, stable power supply in the Washington, DC, area, but it will be nowhere near enough to supply all the electricity needed to get through the winter or summer in that area.

Another truly ambitious hydroelectric energy storage project is on the island of el Hierro, the smallest of the Canary Islands. El Hierro is home to about 11,000 people, including many tourists. The island has a few remaining speakers of a language based on whistling, which shepherds used to communicate with each other when the island was solely agricultural.

The major effort to make the entire island energy self-sufficient uses the tropical sun, the trade winds, and the mountainous geography. Solar cells and wind turbines supply electricity; energy storage uses a half-million cubic meter reservoir on top of a 1500-m mountain. This reservoir stores energy equal to

$$\Delta E = \rho V g h$$
$$= 1000 \text{ kg/m}^3 \left(0.5 \times 10^6 \text{ m}^3\right)\left(9.8 \text{ m/s}^2\right)(1500 \text{ m})$$
$$= 7.4 \text{ TJ}$$

This is enough energy to support the island for a couple of weeks if the wind and solar systems fail. But no one is taking chances: there is a diesel generator, too. And while hydroelectricity is a mature, dependable technology, the huge reservoir is going to require a large capital investment.

7.1.2 Storing Energy as Compressed Gas

Another way of mechanically storing copious amounts of energy is compressing and storing a gas like air. This will work, but the compression of gases normally wastes energy. There are two limiting ways in which gases can be compressed: isothermally or adiabatically, both parts of the Carnot cycle discussed in Chap. 4. Isothermal compression is the most energy efficient: the amount of energy stored can be over 90% of the amount used for the compression. However, compressing a gas isothermally must be slow to ensure near constant temperature. This requires large equipment with abundant heat exchange, which is expensive.

Adiabatic compression is more attractive but less efficient because of entropy constraints, the parallels of those that compromised the Carnot cycle. While the efficiency is theoretically around 70%, the efficiency in practice is closer to 50%, even with cooling after each increment of compression. We can observe the unwanted heating which causes this inefficiency by pumping up a bicycle tire and feeling the pump—it gets hot after the compression. This heat energy is wasted. Although clever ideas of combining compression and absorption can increase efficiency, the potential gains will probably be small.

There are dozens of other approaches that are being developed to store energy mechanically. These include stacking large heavy blocks several meters high to store potential energy, and rotating large flywheels at fast speeds to store kinetic energy. Their costs per MWh of energy stored remain high but are declining.

7.2 Thermal Energy Storage

Another option for energy storage is thermal. The idea is simple—heat or cool a mass, keep it insulated, and use the energy later when it's needed. The greater the mass, the more energy can be stored. Here, we discuss two types of thermal energy storage, sensible and latent. Sensible heat changes the temperature of a mass, like heating cold water to make hot water. Latent heat changes the phase of a mass without changing its temperature, like boiling water to make steam.

7.2.1 Sensible Heat Storage

In Rajasthan, India, where the desert temperatures are high during the day and low at night, houses are made from materials such as brick and sandstone. These materials absorb heat in the day and release it at night, helping to dampen the temperature swings inside the house. The heat capacity of the material is key:

$$Q = m\,C_p \Delta T$$

where Q is the thermal energy stored by a material with mass m, C_p is the material's heat capacity per unit mass, and ΔT is the temperature difference between the mass and the ambient surroundings. Table 7.2 shows values of the heat capacity of several materials. Water has one of the highest heat capacities of any material. We can use these values to estimate the potential of thermal energy storage using sensible heat.

Table 7.2 Heat capacity of materials

Material	Heat capacity	
	kJ/kg-K	MJ/m^3-K
Water	4.2	4.2
Sandstone	0.90	2.2
Brick	0.92	2.0
Concrete	0.92	2.1
Steel	0.48	3.8
Timber	1.2	0.80
Salt (NaCl)	0.86	1.9
Gold	0.13	2.5
Air	0.70	0.0008

Heat capacity of materials can be per unit mass or per unit volume

In the children's fairy tale of the *Three Little Pigs*, we know the smartest pig was the one who built the brick house. After all, his is the only house that the Big Bad Wolf could not blow down. But is the smartest pig also energy conscious? Though we are unsure, we imagine a pig could comfortably live in a single square room of five meters on each side and a two-meter-high ceiling. And we are equally unsure about how hard the wolf huffed and puffed, but we are sure that if the walls, ceiling, and floor of the house were made of just one layer of brick, then the house would stand firm against the wolf's hardest huffing and puffing. We also know that in England, where the story was written, bricks are 215 mm long, 102.5 mm wide, and 65 mm high with a standard 10 mm mortar joint in between layers. The total volume of the bricks to make the pig's house, disregarding doors and windows, is approximately:

$$\text{Volume of bricks} = (4)(5\,\text{m})(2\,\text{m})(0.1025\,\text{m}) \\ + (2)(5\,\text{m})(5\,\text{m})(0.065\,\text{m}) = 7.4\,\text{m}^3$$

where we've assumed the four walls have a thickness of 102.5 mm, the ceiling and the floor each have a thickness of 65 mm, and ignored everything else used to construct the house.

Assuming an average 10 °C day-time temperature difference between the bricks and the ambient surroundings, the thermal energy stored in these bricks is

$$\text{Thermal energy stored} = \left(7.4\,\text{m}^3\right)\left(2.0\,\text{kJ/m}^3\,\text{K}\right)\left(10\,^\circ\text{C}\right) = 148\,\text{kJ}$$

But is this sufficient to warm up the house at night?

Reflection: Estimate the volume of air in the house and use heat capacity of air from Table 7.2. Would 148 kJ of

7 Sustainable Energy Requires Energy Storage

thermal energy be sufficient to warm up the pig's house at night by an average of 15 °C?

Fairy tales aside, a more practical example of thermal energy storage using latent heat is an ordinary hot water tank commonly found in residential homes and commercial buildings. This type of energy storage makes hot water available on demand, not having to wait for the water to heat. Hot water heaters typically store water at approximately 60 °C (140 °F). The thermal energy for heating the water can come from any number of sources—solar, natural gas, coal, electricity, geothermal, or even a heat pump. Natural gas is the most common in the U.S. To reduce carbon dioxide emissions, many communities are providing financial incentives for residents to switch from natural gas-fired water heaters to those using sustainable energy sources, such as high-efficiency heat pumps driven by renewable energy.

Consider a 150 L (40 gallon) water heater. The energy to heat water from 15 to 60 °C is:

$$Q = (150 \, \text{L}) \, (1 \, \text{kg/L}) \, (4.2 \, \text{kJ/kg-K}) \, (60 - 15 \, ^\circ\text{C}) \, (1 \, \text{MJ}/1000 \, \text{kJ})$$
$$= 28 \, \text{MJ}$$

From Table 3.1, natural gas has a heating value of 55.5 MJ/kg, giving us an estimate of the quantity of CO_2 released to heat the 150 L of water:

$$\begin{aligned}CO_2 \text{ released} &= (28 \, \text{MJ}) \, (1000 \, g \, CH_4/55.5 \, \text{MJ}) \\ &\quad (44 \, g \, CO_2/16 \, g \, CH_4) \, (1 \, \text{kg} \, CO_2/1000 \, g \, CO_2) \\ &= 1.4 \, \text{kg}\end{aligned}$$

where we've noted that 44 g CO_2 (1 mol CO_2) results from burning 16 g methane (1 mol methane). Over the course of a year, the average gas-fired home water heater in the U.S. releases 1000 kg CO_2. By switching to a heat pump water heater driven by renewable energy, this 1 tonne of CO_2 emissions can be avoided each year. But remember: in our society, each person has an average CO_2 emission of 16 tons of CO_2 per year.

7.2.2 Latent Heat Storage

Latent heat refers to the energy needed for a material to change phases, from gas to a liquid or from a liquid to a solid. If the material is pure, then the phase transition happens at a single temperature, as when water turns to steam at 100 °C.

To illustrate this, we consider two examples: a cup of coffee and a solar plant. If we pour scalding hot coffee into a mug, we must wait for the coffee to cool down before drinking it. Once it's sufficiently cool, there's a limited time to finish the coffee before it becomes too cold to drink. To give more time to enjoy the coffee, some cleverly designed coffee mugs use thermal energy storage. They have a phase change material inside their walls that transitions from a solid to a gel at about 55 °C, the temperature where most people prefer to drink coffee. The material's transition from a solid to a gel by absorbing heat and from a gel back to a solid by liberating heat. Such mugs therefore have a large built-in 55 °C thermal reservoir, which compared to a regular mug, absorbs heat from scalding hot coffee and releases that heat to keep coffee at 55 °C longer.

A second example of thermal energy storage by latent heat is a concentrated solar power plant. In such plants, vast arrays of parabolic solar reflectors direct the sun's rays towards a central 150 m tall tower. In the tower, the high flux of solar energy melts solid salts into a molten state.

These molten salts are then used both as a heat transfer fluid and a thermal energy storage medium. When the sun is not shining, these molten salts deliver thermal energy at over 560 °C, sufficient to run a conventional steam turbine to generate electricity. Globally, about 7 GW of concentrated solar power plants are installed across the world today, and more are expected in the years. Reducing the costs of such systems is a priority.

7.3 Chemical Energy Storage

We now switch from trying to store energy mechanically or thermally—for example, in a water reservoir or in molten salts—to storing it chemically, as we have done in gasoline. In other words, we are storing energy in some higher chemical energy form, like a fuel. Such storage involves chemical fuels like those shown in Table 7.3. The first column in this table gives four fossil fuels and two sustainable alternatives; the second column gives their energy per mass or specific energy; and the third lists their energy per volume. Energy per mass is an important metric for stationary power plants, and the energy per volume is best for mobile ones, like cars. Note that natural gas, that is, methane (CH_4), has an attractive energy per mass but a poor energy per volume, which is why we heat houses but don't run cars with natural gas. We want systems having both large energies per mass and per volume, but we normally will be forced to compromise.

Chemical energy storage can be much more intense than mechanical energy storage. To support this, we note that 3800 m^3 of gasoline provides the same energy as 34,000,000 m^3 of water held 380 m up in the Bath Energy Storage Facility discussed earlier. The volume of gasoline is about 10,000 times smaller than that of water.

Table 7.3 Comparing common and less common fuels

Source	Specific energy, MJ/kg	Energy density, MJ/L
Natural gas (CH_4)	56	0.04
Gasoline (C_8H_{18})	46	34
Ethanol (CH_3CH_2OH)	30	24
Acetylene (C_2H_2)	50	0.06
Hydrogen (H_2)	120	0.1
Ammonia (NH_3)	23	16

However, generating energy from fuels like gasoline also generates CO_2. To avoid CO_2, many alternatives have been suggested. Three common ones are hydrogen, ammonia, and liquid organic hydrogen carriers (LOHCs), each of which is discussed in the following paragraphs.

7.3.1 Hydrogen

Hydrogen is an attractive fuel, either to burn directly or to use in fuel cells to generate electricity. Burning hydrogen directly is easiest but is subject to the inefficiencies of any heat engine, that is, to Carnot limits. Using hydrogen in a fuel cell avoids the Carnot limits but can require expensive catalysts. Still, both merit consideration.

Hydrogen can be made directly from solar- and wind-generated electricity by the electrolysis of water, a 200-year-old technology optimized within 10% of the maximum efficiency possible. To make the hydrogen, we pass an electric current between two electrodes immersed in water: hydrogen bubbles off one electrode and oxygen off the other. The byproduct oxygen, currently discarded, could be an unrealized opportunity. The overall reaction is:

$$2H_2O(l) \rightarrow 2H_2(g) + O_2(g)$$

where (l) and (g) are the compounds as liquid and gas, respectively. In practice, this overall reaction takes place in two locations. At the positive electrode (the anode):

$$2H_2O(l) \rightarrow O_2(g) + 4H^+(l) + 4e^-$$

where H^+ (l) represents a hydrogen ion—a proton—in water, and e^- is an electron. At the negative electrode (the cathode):

$$4H^+(l) + 4e^- \rightarrow 2H_2(g)$$

The sum of these latter two reactions occurring at different electrodes is the overall reaction given earlier. Thus, electricity supplies the energy to split benign water into two much more reactive gases, oxygen and hydrogen.

Hydrogen can also be made from fossil fuels, especially from methane, the principal component of natural gas. This "steam methane reforming" process was detailed in Sect. 3.4.3, which you may wish to review now. Steam methane reforming does produce substantial amounts of CO_2, like any process based on fossil fuels, but it is currently the cheapest way to make hydrogen. Water electrolysis itself does not make CO_2.

The production of CO_2 has prompted identification of the source of hydrogen by a "color". Of course, hydrogen gas does not have any color. As noted in Sect. 3.4.3, these colors refer to the process used to make the hydrogen—*green* for water electrolysis using renewable energy, *gray* for steam methane reforming, *blue* for steam methane reforming with carbon capture and storage, and *pink* for water electrolysis using electricity from nuclear power. Every few years, we hear of a new method to make hydrogen with a new color assigned to it.

While hydrogen is an excellent fuel, it is difficult to store and can be dangerous. Keeping hydrogen liquid requires low temperatures, about 20 K or − 424 °F. Moreover, while the specific energy produced by mass combustion of hydrogen is the best of any fuel in Table 7.3, the energy per volume is small. We can compress the gas to perhaps 350 bar pressure to improve this energy density, but keeping such a compressed gas in a moving car is a risk that must also be mitigated.

An alternative storage method is to adsorb the hydrogen gas onto a solid, the subject of the enormous research effort summarized in Fig. 7.1 (DOE 2024). This figure plots the overall hydrogen capacity versus the temperature necessary to desorb the hydrogen either for combustion or for use in a fuel cell. The target compounds should have a high capacity and desorb at a moderate temperature, as shown by the dashed box. So far, there is no obvious winner, though the search continues.

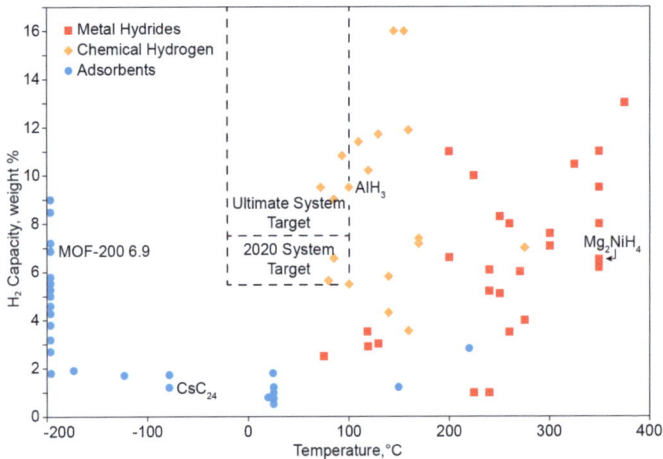

Fig. 7.1 Some Materials Used to Store Hydrogen. The dashed region is the target for commercialization (DOE 2024)

7.3.2 Ammonia

A second sustainable fuel is ammonia. Ammonia, an important chemical in agriculture for over a century, already has an effective infrastructure for its distribution. Its use as a major fuel, or more exactly, as a vector for storing hydrogen, is new.

Ammonia is made by combining nitrogen and hydrogen in what is sometimes described as the most important chemical reaction in the world:

$$N_2(g) + 3H_2(g) \to 2\,NH_3(g)$$

The nitrogen in the reaction comes from air; the hydrogen is currently made largely from natural gas using steam methane reforming. As a result, current ammonia production is an enormous source of greenhouse gases, producing about 3% of the total CO_2 emissions in the world. This is tolerated because ammonia as fertilizer is central to the world's food supply. The nitrogen atoms in our bodies are essential to making proteins and DNA, and about half of the nitrogen in our bodies got there via this reaction. If ammonia synthesis by this reaction ceased, about one billion people would starve.

We are considering ammonia to reduce global warming by using sustainably generated electricity to make hydrogen, which is then used in the reaction above to make ammonia. No CO_2 is emitted, though some nitrogen oxides—"NO_x"—are produced. The ammonia made can be used directly as a fuel, burned in gas turbines. It is especially attractive as a maritime fuel for powering ships. It is ironic that we are considering ammonia for ocean shipping because this is a return to wind power, just as we depended on wind power for clipper ships two hundred years ago.

Ammonia is attractive as a fuel on land because the infrastructure for handling it already exists for agriculture. To be sure, ammonia is a gas under ambient temperature and pressure, but it is easily liquefied and can be stored as a liquid. Such storage already operates safely throughout agricultural communities. At present, ammonia made from fossil fuels arrives via pipelines to large gas storage facilities. These facilities fill so-called "nurse tanks" that are delivered to individual farmers. The farmers then inject ammonia into the soil through the blades of their plows. This distribution system is already operating with an acceptable safety record. Unlike liquid hydrogen, our society has considerable experience handling liquid ammonia.

In addition to using ammonia as a fertilizer, ammonia is a means of storing hydrogen, a "hydrogen vector." In this scenario, we manufacture ammonia in many small plants; ship it to urban areas; decompose it ("crack it") into nitrogen and hydrogen; and feed the hydrogen to a fuel cell to make electricity. In other words, we are converting the ammonia back into nitrogen and hydrogen to be used as fuel. It is also possible to burn the ammonia directly in a turbine. Other compounds for storing hydrogen, shown in Fig. 7.1, have not been developed to the same degree as ammonia.

7.3.3 Liquid Organic Hydrogen Carriers (LOHCs)

A different route for storing hydrogen converts hydrogen into a liquid organic compound, which can be transported easily. In one example, hydrogen made with solar power in the Bavarian Alps is reacted to make an organic liquid, which is trucked to industrial Hamburg in northern

Germany. The hydrogen is reacted and released for industrial use, and the trucks return with the recovered organic liquid to Bavaria.

This is an attractive idea that builds tightly upon existing infrastructure. We only must find a chemical carrier for hydrogen that can be reacted with hydrogen produced by electrolysis from water. The reaction yields a liquid which is easily transported and at the destination decomposed to recover the hydrogen. Possible systems include toluene, n-ethyl carbazole, formaldehyde, and dibenzyl toluene. However, no compelling liquid example has yet been identified.

7.4 Electrochemical Energy Storage

Another way in which we can store chemical energy is electrochemical energy storage, that is, batteries. Several types of batteries are possible, as shown in Table 7.4. Batteries have a greater energy density than mechanical methods of storing energy, like water reservoirs or compressed gases, but have a lower energy density than liquid fuels. However, common fuels like gasoline produce CO_2, which we want to avoid. Hydrogen as fuel burns to water vapor. It is easily made from sustainable sources, but it is explosive and difficult to handle. Ammonia as fuel does have major potential, but it is less developed. The convenience of batteries may overcome their smaller specific energies.

Batteries are less complicated physically than the engines or turbines used to burn fuels. They can be smaller and lighter and require less maintenance. In the following paragraphs, we focus on three types of batteries with high short-term applications in energy storage. The first is the

Table 7.4 Batteries for energy storage

Battery	Specific energy, MJ/kg	Energy density, MJ/L
Lead acid	0.12	0.25
Lithium ion	0.7	2.0
Vanadium redox	0.07	0.06
Alkaline	0.5	1.3

Several types of batteries are possible

lithium-ion battery, the workhorse of electric automobiles. Second, we describe the lead-acid battery, the standard for the previous century. Finally, we will discuss the vanadium redox battery. While this redox battery has a lower specific energy than the other types, it is especially suited for stationary energy storage produced from solar cells and wind turbines. As the table shows, the alkaline batteries used in flashlights and smoke detectors have excellent properties; we do not discuss these here because they are too expensive for large-scale energy storage.

At the same time, we should remember one other feature we want in batteries: they should be quick to charge and discharge. In many cases, the speed of charging and discharging is slower than we would like. This is especially true of batteries based on solid electrodes, including both lithium-ion and lead-acid batteries. In these systems, we are normally inserting charged species into a solid. For example, in lithium-ion batteries, we are inserting lithium ions into either cobalt or graphite electrodes. This insertion includes the slow diffusion of ions into and out of a solid. To accelerate this diffusion, we often will try to make the electrodes porous so that we can move the ions relatively quickly through liquid-filled pores rather than through solids. While these porous electrodes can work effectively, they are often less stable, so the electrodes function effectively for fewer cycles. This makes us consider batteries which have a lower energy per volume but which store ionic reactants in solution, and hence run for many more cycles of charging and discharging. The balance between high energy storage density and long battery life is an unresolved problem in energy storage.

7.4.1 Lithium-Ion Batteries

The biggest advance in battery technology in the last 20 years has been the development of the lithium-ion battery, key to electric automobiles. In this battery, lithium replaces lead, so the battery is much lighter. It is still heavy. In a typical electrical automotive, the battery weighs 625 kg with a volume of 0.4 m^3. This is a volume of 100 gallons or of a queen-sized bed mattress, a large volume in a passenger car.

The basic structure of the battery, shown in Fig. 7.2, has chemical reactions that are not always cleanly stoichiometric. The key reactions involve an intercalated lithium-carbon complex, an organic lithium electrolyte, and a lithium cobalt salt. The overall reaction is:

$$LiC_6(s) + CoO_2(s) \rightarrow C_6(s) + LiCoO_2(s)$$

where (s) indicates a solid material. The individual reaction within the solid cathode is:

$$CoO_2(s) + Li^+(l) + e^- \rightarrow LiCoO_2(s)$$

That within the solid anode is:

$$LiC_6(s) \rightarrow C_6(s) + Li^+(l) + e^-$$

where (l) refers to a species in the liquid electrolyte. The electrons produced are conducted via solid carbon

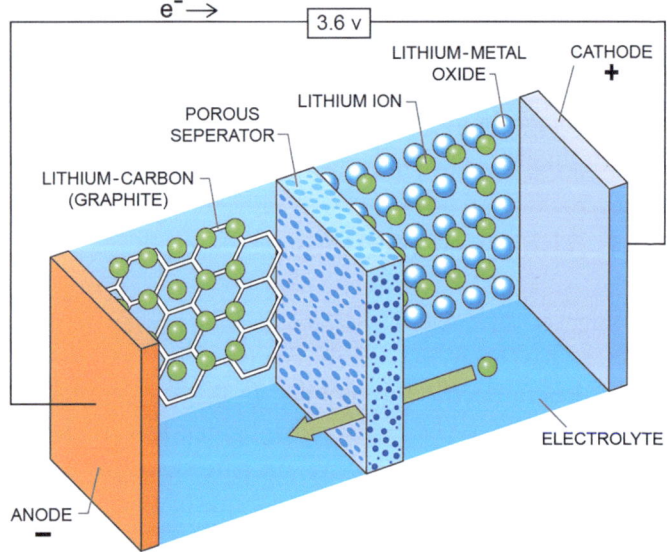

Fig. 7.2 The basic structure of a lithium-ion battery. The energy comes from lithium-ion energy reduction

and cathode salts through an external circuit. The rate-determining step of the lithium transport determines how fast the battery can be charged and discharged. Because lithium transport in solids can be slow, especially across any solid electrolyte interface (SEI), this transport is a focus of battery improvements.

Lithium batteries have an energy per mass six times larger than a lead-acid battery; and the energy per volume is ten times greater. This higher energy is the driving force toward electric car manufacture. The large energy per volume is not surprising; while the weight of the battery case and wiring will be similar in lithium and lead-based batteries, the molecular weight of lithium to lead is 7 to 207, which makes lithium battery energy density higher. The lithium-ion battery is new enough so that we can anticipate significant future improvements.

7.4.2 Lead-Acid Batteries

We next summarize the lead-acid battery, a standard method of electrochemical energy storage since 1859. Here, the total reaction can be written:

$$Pb(s) + PbO_2(s) + 2H_2SO_4(l) \rightarrow 2PbSO_4(s) + 2H_2O(l)$$

where (s) and (l) indicate materials that are solid or in liquid aqueous solution, respectively. At the solid negative electrode, the reaction during battery discharge is:

$$Pb(s) + HSO_4^-(l) \rightarrow PbSO_4(s) + H^+(l) + 2e^-$$

At the solid positive electrode, the reaction is:

$$PbO_2(s) + HSO_4^-(l) + 3H^+(l) + 2e^- \rightarrow PbSO_4(s) + 2H_2O(l)$$

The sum of these two reactions is that given at the start of this paragraph.

The battery itself is shown schematically in Fig. 7.3. The individual reactions given are for discharge, when we are removing electrons to do work. If we are charging the battery, the directions of the reactions are reversed, and we must add energy to the system as electrons.

Like the lithium-ion battery, the lead-acid battery must be recharged slowly to avoid compromising the battery structure. In the lithium case, this was caused by altered interfaces; in the lead-acid case, it is a result of side reactions forming lead sulfate. In both cases, fast charging and repeated cycling comprise performance. This is less true

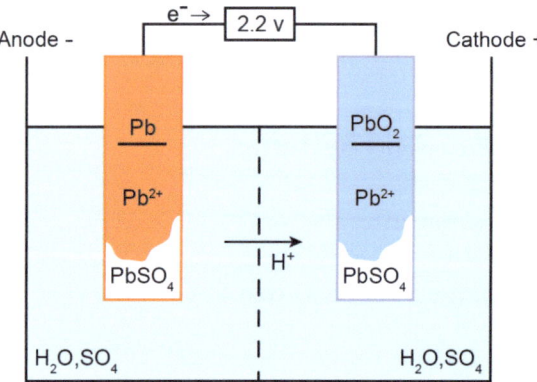

Fig. 7.3 A schematic diagram of a lead-acid battery drawing current. During discharge, the lead is converted to a lower energy. During charging, the flow of electrons is reversed

of the lower energy densities of redox batteries, which are discussed next.

Reflection: What is the history of the lead-acid battery? What technical changes made it better?

7.4.3 Vanadium Redox Batteries

The third battery we want to summarize is not like the lithium-ion, where the lithium ion is put in and taken out of graphitic carbon, nor like the lead acid, where the electrochemical reaction involves solid lead salts. Each of these cases has a higher energy density than redox alternatives like those based on vanadium. However, each of these cases involves slow diffusion in and out of solids, which means these batteries are difficult to charge and discharge quickly.

Batteries based on vanadium are different, as shown in the schematic diagram in Fig. 7.4. They use two different

7 Sustainable Energy Requires Energy Storage

vanadium-containing aqueous solutions separated by a membrane highly permeable to protons (H⁺) but impermeable to anything else present. The membrane used, often an ionic form of the perfluorinated polymer Teflon, was developed for making chlorine gas. The overall reaction during discharge for this battery is:

$$VO_2^+(l) + 2H^+(l) + V^{2+}(l) \rightarrow V^{3+}(l) + VO^{2+}(l) + H_2O$$

where again (l) is a species in liquid (aqueous) solution. More specifically, at the surface of the anode, the reaction is:

$$V^{2+}(l) \rightarrow V^{3+}(l) + e^-(l)$$

At the cathode, the reaction is

$$VO_2^+(l) + 2H^+(l) + e^- \rightarrow VO^{2+}(l) + H_2O$$

The sum of these two half reactions is the overall reaction shown above.

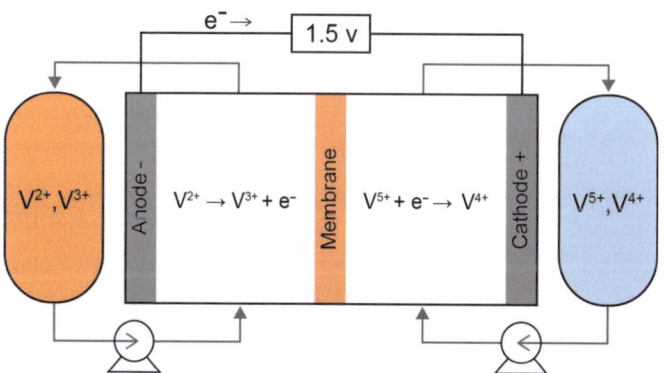

Fig. 7.4 A vanadium redox battery. It has a lower energy density but charges and discharges faster because it does not involve diffusion in solids

If we look in Table 7.4 for the energy density supplied by these three batteries, we see the vanadium example has less than 10% of the energy density supplied by the lithium battery described above. Somewhat surprisingly, the vanadium battery is more attractive for seasonal energy storage. It is cheaper to build and expand, and it responds rapidly electrically. If we based an automobile on this type of battery, it would need a gas tank 10 times the size of a regular gas tank—but it could be recharged in a couple of minutes. While vanadium batteries seem a poor bet for electric cars, they are a promising option for damping out the cycles in wind and solar power generation. This battery has potential for energy storage.

7.5 Considerations for Energy Storage

Energy storage is key for enabling large-scale sustainable energy. Two considerations that impact the potential of any energy storage approach are the round-trip efficiency and the cost of the technology. The first consideration, round-trip efficiency, is defined similarly to the other efficiencies we defined in Chap. 4: energy we want divided by energy that we pay. Large-scale hydroelectric energy storage has round-trip efficiency of about 80%. If we use 100 MWh of electricity to pump water from the bottom of a dam top and let the water run back down again through a turbine, the turbine will generate about 80 MWh of electricity. The remaining 20 MWh is lost to heat generated somewhere in the system—friction between moving parts, electrical resistance, and the like, all of which generate entropy. Remember that the second law of thermodynamics states that entropy of the universe is increasing,

7 Sustainable Energy Requires Energy Storage

Table 7.5 Round-trip efficiency

Storage technology	Round-trip efficiency (%)
Hydroelectric	70–80
Flywheels	80–90
Batteries	75–90
Compressed air	65–75

meaning that round-trip efficiency will always be less than 100%. Table 7.5 shows the round-trip efficiency of several energy storage technologies at commercial scales.

Of course, the round-trip efficiency is also dependent on the time needed to store the energy. A flywheel cannot be left spinning indefinitely; friction will gradually convert its kinetic energy to heat. A commercial-scale battery bank loses about 2% of its stored energy each month to ion diffusion and subsequent reactions occurring within the battery. The selection of a commercial-scale energy storage technology must consider these factors as well as cost, capacity, energy density, and the like. The values in Table 7.5 therefore should be considered in the context of how the technology is being used.

The second consideration of energy storage is cost, which, of course, is not unique to energy storage, but it does pose interesting limits. In a sustainable society, energy storage is needed not only across short time scales such as hours and days, but it is also needed across longer time scales such as the months between summer and winter.

An energy storage system can be charged during times with excess renewable energy and discharged during times when renewables are offline. Such a system can potentially be all renewable energy. But the cost of the energy storage system must be such that it can economically operate between the difference in energy prices during low and high demand. For short-duration needs, such price differences can occur hundreds of times each year, offering ample opportunity to amortize the energy storage system cost over its lifetime.

Long-duration energy storage is particularly challenging because not only must the system store energy for months with minimal energy leakage, but it must also be cheap enough that it can be cost-effective by charging during the summer and discharging during the winter. This annual charging and discharging cycle offers far less opportunity to amortize the energy storage system cost over its lifetime. Cost-effective, large-scale long-duration energy storage is particularly challenging, though several promising approaches are being pursued, primarily based on chemical energy storage.

7.6 Conclusions

The chapter asserts that energy storage will be essential for a sustainable energy system. Wind and sun are certainly going to be major sources of this energy, and they are obviously periodic. We need to store energy for times when the sun is not shining or the wind is not blowing. We need to store energy over a few days, between day and night. We also need larger energy storage over seasons, between winter and summer. At the moment, energy storage over daily variations is being commercially solved by several technologies, particularly batteries. Storage of potential energy as changes in water height is reliable and well developed, though probably not large enough to handle seasonal fluctuations. Storage of energy as compressed air is attractive, though compromised by adiabatic inefficiency. And there are dozens of other approaches being developed that include gravitational, mechanical, and thermochemical means. Still, handling these daily variations looks reasonable, but energy storage costs need to be driven down further.

Seasonal variations are more demanding. They will likely be handled by chemical energy storage. Making hydrogen from water electrolysis using renewable energy is one such idea. The hydrogen can be stored indefinitely until needed. However, because hydrogen is a gas, it has relatively low energy density by volume even when it's compressed. There are also safety considerations with storing large amounts of hydrogen. As a result, other chemicals are being considered, such as ammonia. Ammonia can be compressed to a liquid, providing much higher energy density. But the process to make ammonia is itself energy intensive, and the costs are high, making it challenging to use for long-duration energy storage.

Sustainable energy may be cheap when available, but cost-effectively mitigating the effects of its periodicity is a major technical challenge.

Reference

Energy Efficiency and Renewable Energy, US Department of Energy (DOE). 2024. *Materials-Based Hydrogen Storage*. Accessed 1 Dec. 2024. https://www.energy.gov/eere/fuelcells/materials-based-hydrogen-storage.

8

Carbon Capture and Storage

Our society demands energy when and where we need it. We expect our lights to turn on when we flip a switch, our cars to operate whenever we want, and our factories to make products year-round. For an energy producer, this means some of the energy should be generated all the time, called "baseload," while some of the energy should be generated on demand, called "dispatchable". Now we also want this energy to be decarbonized. As increasing amounts of renewable energy enter the energy mix, its variability poses increasing challenges to ensure power is available when needed.

There are only a few low-carbon options that can provide decarbonized and dispatchable electricity —renewables with energy storage, nuclear power, and fossil fuels with carbon capture and storage (CCS). We've covered renewables and energy storage throughout this book, briefly touched nuclear power in Sect. 5.4, and we focus on CCS in this chapter.

8.1 California's Duck Curve

California has some of the highest renewable energy generation in the U.S. Figure 8.1 shows the "Duck Curve", a plot of the electricity supplied over the course of a typical spring day by non-renewable sources like natural gas and nuclear (U.S. Energy Information Administration 2023). During daylight hours between 8 am and 6 pm, when California bathes in sunshine, solar power increases and non-renewable power decreases. At some portions of the day, solar and wind provide all the power needed by the state. It's remarkable. However, around 6 pm when solar power decreases during evening and night or is otherwise unavailable, traditional power plants must steeply ramp up to meet power demand and ensure the power grid is stable and reliable. This rapid ramping up and down puts strain on a traditional plant's machinery and increases its operational cost since their power is sold only for limited time periods. Note that adding additional solar power results in these curves becoming steeper, morphing the "duck" into a "canyon", and that each additional solar unit has diminishing impact. Indeed, in California, brief periods of overgeneration of wind and solar have caused electricity prices to become negative during the day.

The seemingly obvious solution to this overgeneration problem and power grid management is energy storage, which we covered in Chap. 7. An energy storage system can be charged during the day with excess renewable energy and discharged at night when renewables are offline. Such a system can potentially be completely based on renewable energy. But the cost of the energy storage system must be such that it can economically operate between the difference in energy prices across the day. And short-duration energy storage systems across the day are not necessarily suited for long-duration energy storage across

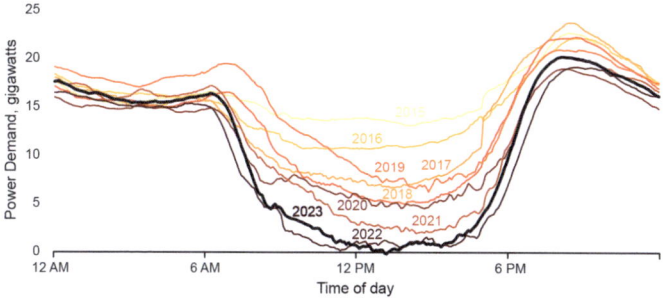

Fig. 8.1 California's Duck Curve. The amount of power generated by non-renewable sources drops to near zero during daylight hours and curves steeply at dawn and dusk (U.S. Energy Information Administration 2023)

summer and winter months. Often, maintaining power reliability requires additional generation capacity to serve as backup, adding additional cost to an overall energy system.

Nuclear power and coal-fired power plants do not ramp up and down quickly, making them challenging to operate under demand-following conditions like steep duck curves: they are more suitable for baseload power. Natural gas-fired power plants can ramp up and down quickly to supply the power needed, but they release CO_2.

Another option to provide system stability and power reliability involves managing the energy demand. Sometimes, during extremes in weather, the power demand is much higher than the energy system can deliver, and the authorities sometimes plead for citizens to reduce their power consumption. In some cases, citizens cede control of some home appliances, such as water heaters and air conditioners, to system operators who can alter the citizens' energy consumption when needed. Such control requires the power grid to be modernized and internet-connected with the appliances. In Palo Alto, California, some of these changes are already underway.

In addition to these present challenges, electricity consumption in the coming years is projected to accelerate because of electric vehicles, the electrification of industries, and power-hungry data centers for the internet and artificial intelligence. As our energy system shifts more towards electricity with renewables, grid operators will face increasing challenges to ensure reliable, low-carbon, dispatchable power. As the California Duck Curve illustrates, there is no single solution to address all these issues. A portfolio of technology options will be needed in the energy transition.

8.2 Carbon Capture and Storage (CCS)

All these low-carbon dispatchable power generation options are capital-intensive and need to operate for decades to make them cost-effective. The less any unit or system operates, the more expensive it becomes per unit of energy delivered. Many power companies have resource planning departments that project future power demand and how to meet that demand at the lowest cost. Often, the lowest cost option with near-complete decarbonized energy shows some amount of CCS. Globally, CCS is estimated to be needed for 8% of anthropogenic emissions for the least-cost pathway to decarbonization by 2050 (International Energy Agency 2023). These models show that if we remove CCS as an option, global decarbonization costs will significantly increase.

Carbon capture and storage refers to separating carbon dioxide from a gaseous mixture and injecting it underground, where it's permanently trapped and stored. The gas mixture to be treated by CCS is often the flue gas from combustion of fossil fuels in air, such as a power plant, a cement plant, or an industrial boiler. CCS can be applied to separating carbon dioxide from the atmosphere too, which is often called "direct air capture."

CCS is controversial because capturing and sequestering the carbon dioxide allows fossil fuels to continue to be an energy source, albeit a decarbonized one. Some see it as "kicking the can down the road." After all, they argue, if CO_2 from fossil fuels is the problem, then just get rid of fossil fuels. But fossil fuels supply more than 80% of the global energy, a number that has not changed for decades. The *growth* in global energy consumption continues to be dominated by fossil fuels, resulting in increasing fossil energy consumption and increasing CO_2 emissions practically each year. Moreover, global energy and climate models show that we cannot make the transition from fossil fuels to renewables without considerable cost. CCS reduces this cost. Indeed, CCS may even play a role in removing carbon dioxide from the atmosphere to recover a stable climate for the future. We acknowledge that CCS appears to be conducting business as usual, but we have come to recognize that fossil fuels are highly unlikely to be displaced in the timeframes needed to mitigate climate change. CCS seems a necessary medium-term salve.

The process of CCS is a series of steps—capture, compression, transport, and storage of the CO_2. These steps, sketched in Sect. 2.4, are discussed further in this chapter, with the potential role of CCS becoming more specific in Chap. 9.

8.2.1 Carbon Capture

Imagine we have a gaseous mixture that contains carbon dioxide. Such gases typically result from combustion of fossil fuels in air to release energy in power plants, steel manufacturing facilities, and cement kilns. We want to separate the CO_2 and prevent its release into the atmosphere.

The first step, called "carbon capture", involves separating the carbon dioxide from the rest of the gaseous mixture, in effect making pure CO_2. There are many approaches to doing so. The most common is to contact the gaseous mixture with a solvent like aqueous monoethanolamine that chemically reacts with CO_2, as shown in Fig. 8.2. The CO_2-containing gas flows upwards in an absorber where it contacts a downward-flowing solvent that reacts preferentially with CO_2. Because the absorber column is often designed tall enough to remove nearly all the CO_2, the remaining mostly CO_2-free gas can be vented to the atmosphere. The CO_2-loaded solvent, which leaves the bottom of the absorber column, is sent to a second column, called the stripper, where the solvent is heated to release the CO_2. The CO_2 is cooled, dehydrated, and compressed to produce a supercritical fluid for storage. The regenerated solvent flows back to the top of the absorber. The solvent therefore flows in a loop, capturing CO_2 in the absorber and releasing it in the stripper.

Other approaches to capturing CO_2 include contacting the gas mixture with solid particles that selectively adsorb CO_2, using a membrane that selectively permeates CO_2, or cooling the gas such that the CO_2 precipitates to become dry ice. These approaches are less common than gas absorption, but they are making advances through research.

All these approaches require energy to separate the CO_2. This energy can be heat for solvents or solids to release the captured CO_2, vacuum or pressure for membranes to permeate CO_2, or a refrigeration cycle for cryogenic capture. Of course, in addition to energy, carbon

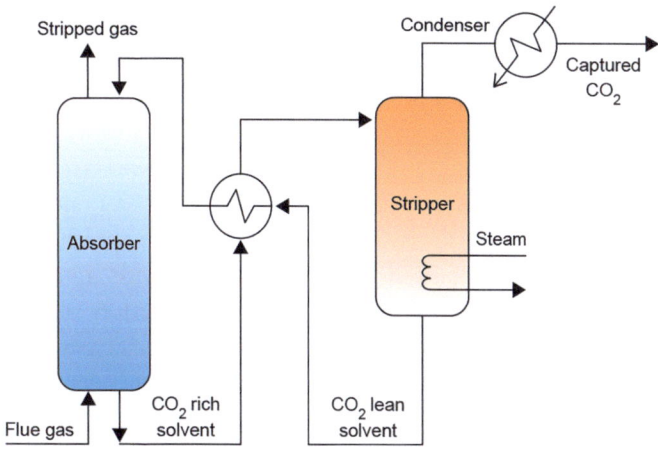

Fig. 8.2 Carbon capture. A solvent circulates between two columns, capturing CO_2 in the absorber column and releasing it in the stripper column

capture technologies require hardware too—columns, heat exchangers, pumps, compressors, ducts, and piping. Both energy and hardware add cost relative to what we do today, which is nothing. Today, we release the gas mixture containing CO_2 directly into the atmosphere. In the absence of regulatory or economic drivers to stop such CO_2 emissions into the atmosphere, there simply is no incentive to add the cost of CCS to a process.

The energy required to drive the carbon capture process must at least be equal to or greater than the thermodynamic minimum work that we derived in Sect. 4.3. Of course, practical systems will require more than this minimum. In Chap. 4, we calculated the theoretical minimum work needed to conduct a separation of CO_2 from a gas mixture:

$$\Delta G(\text{unmix})/nRT = -x_1 \ln x_1 - (1-x_1) \ln(1-x_1) > 0$$

where ΔG is the minimum work needed; n is the number of moles; R is the gas constant; T is the temperature; and x_1 is the mole fraction of CO_2 in the mixture. Figure 8.2 shows this minimum work for each tonne CO_2 captured as a function of fraction CO_2 in the gas mixture we start with, as well as the minimum work of 68.8 kWh/t CO_2 needed to compress CO_2 from atmospheric pressure to 150 bar. Compression is required to prepare the CO_2 for transport and storage, discussed further in the next section. This is the minimum; the actual energy required will depend on the capture technology, the hardware design, and eventually the economics of the process.

How we supply the energy for capture and compression is a key consideration. If the energy comes from renewables or nuclear, then it often makes more sense to put this energy source onto the power grid to replace fossil fuels and avoid CO_2 emissions in the first place. As a result, carbon capture often relies on energy from the fossil fuel itself. In other words, fossil fuels are used to decarbonize fossil fuels.

8.2.2 CO$_2$ Storage

Once CO_2 has been captured, the only large-scale option to manage the billions of tonnes of CO_2 emitted annually is to store the CO_2 underground. In this process, CO_2 is compressed into a liquid-like state—a supercritical fluid—transported typically by pipeline to a suitable geological formation where it's injected approximately 2–3 km underground, as shown in Fig. 8.4 (CO2CRC 2017; U.S. CBO 2023). At these depths, CO_2 remains a supercritical fluid, permeates porous rocks and fissures, reacts with other minerals, and is effectively trapped over geological time frames. There are four major mechanisms of trapping CO_2. "Structural trapping" is the physical trapping of CO_2 in the reservoir. Because CO_2 is more buoyant than other liquids in the pore spaces, it will migrate upwards towards the impermeable caprock. "Residual trapping" refers to small portions of CO_2 left in pores as the CO_2 migrates through the porous rocks. The porous rocks act just like a porous sponge for CO_2. "Solubility trapping" refers to the physical dissolution of CO_2 into the brine water that is present in the rock formation. This dissolution occurs at the CO_2-brine interface. "Mineral trapping" refers to CO_2 reacting with brine and other minerals in the rock to form solid carbonates such as $MgCO_3$. These reactions occur since CO_2 and water react to become carbonic acid (H_2CO_3) and bicarbonate (HCO_3^-) which undergo slow reactions with minerals, permanently trapping the injected CO_2. These four mechanisms help trap and store CO_2 in suitable geology.

Not all geological sites are suitable for CO_2 storage. A thick impermeable caprock above the formation must be present to ensure CO_2 does not permeate back to the surface. Below the caprock, rock porosity, rock permeability, and average pore size of the formation influence the amount of CO_2 that can be stored. Another factor is the distance from the CO_2 capture site to the storage reservoir. Pipelines can transport large quantities of CO_2 from the capture site to the storage site; ships, rail, and trucks are options for smaller amounts.

Fortunately, the U.S. has potential geological sites for thousands of gigatonnes of CO_2, enough to store CO_2 emissions for the next several hundred years. But this geology is not uniformly distributed across the country, and thus, detailed site characterization must be done before a particular site can be deemed suitable for CO_2 storage. Many other parts of the world also have storage potential, but these sites are not uniformly distributed.

8.3 Where CCS Has Promise

We often focus on power plants because they generate about a third of all global CO_2 emissions, the most of any sector, and they are society's key machines that convert heat to work. Because power plants are large stationary sources of CO_2 emissions, CCS could be commercially viable because of economies of scale. Transportation using trucks, planes, and planes is the next major source of CO_2, about a sixth of global emissions. Because they are small mobile sources of CO_2 emissions, CCS is more challenging, and other decarbonization options may be lower cost.

Capturing CO_2 from a power plant is straightforward, but expensive. A coal plant is typically around 35% efficient overall. In other words, for 1 MW_t thermal energy input, a plant produces 0.35 MW_e electrical energy output. In Chap. 1, we noted that coal releases about 90 kg CO_2/ 10^9 J of thermal energy. Therefore, we can estimate that for each MW_e of power, a coal plant releases about 0.9 t CO_2 per hour:

$$\left(0.09 \text{t} CO_2/10^3 \text{ MJ}\right)(\text{MJ}/MW_t - s)(MW_t/0.35 \, MW_e)(3600 \, s$$
$$= 0.9 \text{t} CO_2/MW_e - h$$

We now want to estimate the amount of parasitic loss a CCS process would impose on such a plant. To do so, we note that flue gas from a coal power plant has between 11 and 14% CO_2. As shown in Fig. 8.3, the minimum thermodynamic work to capture and compress all CO_2 from such a gas stream is approximately 130 kWh/t CO_2. This value is about 11% of the net output of the power plant (1 MWh/0.9 t CO_2). A practical carbon capture and compression process available today needs about twice this minimum work, or about 20–22% of the net output of the power plant. This is often called the "energy penalty" or "parasitic load" of CCS on a coal-fired power plant.

A similar calculation can be made for a natural gas-fired combined cycle (NGCC) power plant whose overall efficiency is about 45%. And in Chap. 1, we noted that natural gas releases about 50 kg $CO_2/10^9$ J of energy. Hence, for each MW_e of power, an NGCC plant releases about 0.4 t CO_2 per hour:

$$\left(0.05\ tCO_2/10^3\ MJ\right)(MJ/MW_t - s)(MW_t/0.45\ MW_e)(3600\ s/h)$$
$$= 0.4 tCO_2/MW_e - h$$

For the same power output, NGCC power plants release about half the amount of CO_2 relative to coal. Flue gas from an NGCC power plant contains 3–4% CO_2, which results in a minimum thermodynamic work to capture and

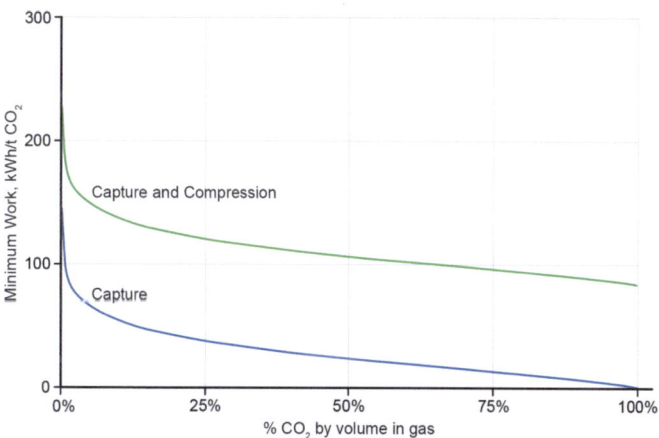

Fig. 8.3 Minimum thermodynamic work for capturing CO_2 at 40 °C and compressing it to 150 bar. Separations of more dilute mixtures of CO_2 cost more energy per mass of CO_2

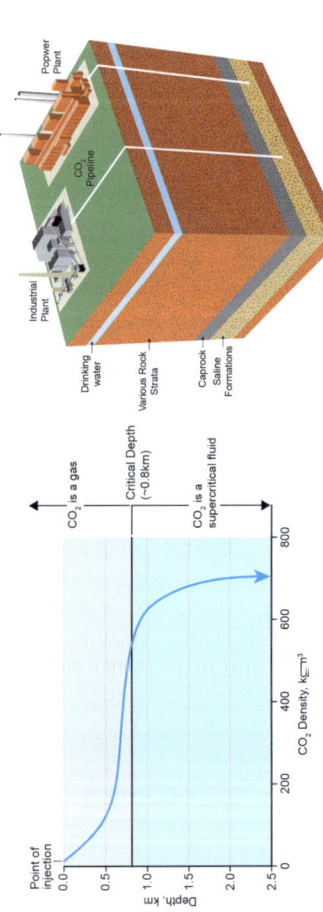

Fig. 8.4 CO₂ storage. a CO_2 is compressed to a high-density supercritical fluid and **b** stored underground in suitable geological formations (based on CO2CRC 2017; US Congressional Budget Office 2023)

compress all CO_2 of about 155 kWh/t CO_2. This value is about 6% of the net output of the power plant (1 MWh/ 0.4 t CO_2). A practical carbon capture and compression process available today is about twice this minimum work, and thus the "energy penalty" of CCS for NGCC is about 10–12%.

Though power plants can vary, these values of 0.9 t CO_2/ MW_e-h for coal and 0.4 t CO_2/ MW_e-h for NGCC provide reasonable estimates of CO_2 emissions from coal and NGCC power plants. CCS on these plants imposes a 20–22% energy penalty for coal and 10–12% energy penalty for NGCC. Transport and storage have less energy consumption compared to the separation costs. CCS will work, but at some cost.

Reflection: Calculate the minimum work needed to capture CO_2 from a steel mill, which typically releases 10–30% CO_2, and a cement kiln, which typically releases 15–30% CO_2.

8.4 Direct Air Capture

As the world continues to emit more CO_2, it's now inventible that emissions are going to exceed the carbon budgets we have left to minimize the impact of climate change. Decades of inaction have gotten us to this point. One proposed approach to addressing this issue is to exceed the carbon budgets now, with the hopes of removing CO_2 from the atmosphere later. This approach is broadly called carbon dioxide removal or negative emissions.

One example is to use CCS on the atmosphere itself, often called direct air capture (DAC). The CO_2 concentration in the air today is about 425 ppm or 0.0425%, rising 2–3 ppm each year. This concentration is about three hundred times more dilute than flue gas from a coal-fired power plant and one hundred times more dilute than flue gas from a natural gas-fired power plant. As shown in Fig. 8.3, the minimum thermodynamic work for capturing CO_2 at such low concentrations increases sharply. If we capture all CO_2 from a given volume of air, the thermodynamic minimum work for capture and compression is about 213 kWh/t CO_2. A practical process for such dilute CO_2 will consume around five times more than this minimum, a value that approaches the output of a coal or NGCC power plant. Hence, the substantial energy needed to drive DAC most often will come from renewable or nuclear energy.

Another challenge for DAC is that for each tonne of CO_2 captured, DAC must process much larger volumes of gas relative to more concentrated flue gas sources such as coal and NGCC power plants. As a result, DAC processing equipment is very large. Removing 1 Mt CO_2/year from the atmosphere will require hardware on the order of 10 m tall and 5 km long. The combination of high energy and high capital cost makes DAC an expensive technology. Still, because humanity is on track to exceed carbon budgets, DAC and other carbon dioxide removal technologies are likely going to be required if climate goals are going to be achieved. This will be expensive, and results directly from society's inaction to reduce emissions earlier. One way or another, we are going to pay a high price for kicking the can down the road.

Reflection: Trees can remove CO_2 from the atmosphere. Can enough trees be planted to significantly impact climate change? What factors should be considered before embarking on such a mission?

8.5 Conclusions

Fossil fuel power plants generate enormous amounts of CO_2. A 500 MWe coal-fired power plant operating 90% of the time during the year generates about 3.5 million tonnes (Mt) of CO_2 each year, while a gas-fired combined cycle plant generates about half as much. While renewable energy is often cheaper, it varies during the day, week, and season. In contrast, a coal or gas-fired power plant can operate as a baseload, and CCS offers a way to decarbonize it. As more renewable power enters the grid, non-renewable power plants are operating in a cyclic fashion, generating power when renewable energy is unavailable. This flexible operation of a power plant is one of the challenges power companies are facing while ensuring the reliability of the overall power grid in a cost-effective manner. CCS will be needed if the energy transition is to be at the lowest cost.

References

CO2CRC Storage and Capture Demonstration. 2017. *Presentation at the Carbon Management Technology Conference*, 19 July 2017.

U.S. Energy Information Administration. 2023. *As Solar Capacity Grows, Duck Curves Are Getting Deeper in California*. Accessed 1 Dec 2025. https://www.eia.gov/todayinenergy/detail.php?id=56880.

U.S. Congressional Budget Office. 2023. *Carbon Capture and Storage in the United States*. Accessed 1 Dec 2025. https://www.cbo.gov/publication/59832.

9

The Energy Transition

Energy development has driven the industrial revolution and is responsible for the structure of our current society. For over three hundred years, this development has been dominated by fossil fuels. First, coal was most important; later, oil became predominant; and today, natural gas is increasingly used. However, even though fossil fuels provide over 80% of the world's energy, their future is uncertain. While there appears to be centuries of coal left, coal combustion causes considerable pollution, especially greenhouse gases like carbon dioxide. The supply of oil is also finite, but oil production is still expected to grow in the next decades. Because fracking has reduced the cost of natural gas, its development is more attractive, but natural gas still causes half the CO_2 emissions of other fossil fuels. Fugitive emissions, that is, accidental escape of natural gas, are also a significant factor in global warming. While the lifetime of methane in the atmosphere is only about

12 years compared CO_2 of 120 years, methane absorbs heat far more than CO_2. Consequently, methane is responsible for around a third of the current rise in global temperature.

Over the long term, the future evolution of society will depend on sustainable, decarbonized energy and new technologies, including potentially nuclear fusion. As the earlier chapters of this book show, this sustainable energy will be dominated by the sun and wind. The sun's energy will be collected with photoelectric cells based on silicon. Wind energy will use turbines which are already well developed. The storage of energy from these periodic resources is a challenge, forcing us to consider alternative technologies like carbon capture and storage.

The energy transition to a decarbonized future has already started, though it will take decades more with potential societal adjustments. But climate change is now upon us, and it will cause even more social disruption unless we act hastily. We have explored earlier in this book the technologies through which this transition can occur.

In this final chapter, we want to outline actions that our entire society could consider. These actions hinge on three questions:

1. Is climate change real?
2. What can we do about it now?
3. How fast can we act over longer times?

We begin to frame answers to these questions in this chapter. The details of the answers will emerge only as the political and societal will to act develops.

9.1 Is Climate Change Real?

We recognize that judgments about climate change, and specifically global warming, come from two different communities: academia and society. Both communities agree that global warming is occurring: the chemistry connecting increased carbon dioxide concentration and increased temperature is established in detail and rarely challenged. The academics are almost completely convinced that increased carbon dioxide comes from human activity. Some of the general population hope that increased carbon dioxide comes from short-term fluctuations, not from human activity. They assert that we have made good CO_2 measurements only for a century or so, and the apparent increase may soon slow.

Actually, we do have measurements of CO_2 for over 800,000 years, which are inferred from ice cores of glaciers. These ice core measurements agree closely with direct measurements for the last century, as shown in Fig. 9.1 (NOAA 2024a). These data show that even through earth's warm periods and ice ages, carbon dioxide concentrations have not risen beyond 300 ppm. In fact, for the last 1000 years, the carbon dioxide concentration has been nearly constant at about 280 parts per million (ppm) until the end of the eighteenth century, and then showed a sharp rise to the current value of over 425 ppm, rising 2–3 ppm each year. This rise is one part of the evidence of human-caused global warming.

While global warming from increasing CO_2 is one part of climate change, the other part is ocean acidification. About 30% of CO_2 emitted into the atmosphere is absorbed into the oceans and reacts to form carbonic acid (H_2CO_3), bicarbonate (HCO_3^-), and other chemicals. The pH of the ocean's surface waters today is about 8.1, a drop of about 0.1 since the start of the industrial

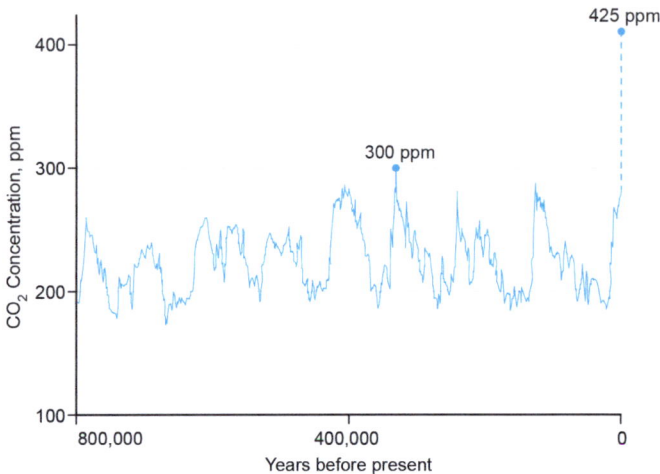

Fig. 9.1 Carbon dioxide concentrations for the last 800,000 years (NOAA 2024a). The recent increase in CO_2 concentration is a result from human activity

revolution (NOAA 2024b). Because pH is a logarithmic scale, this means the ocean surface waters are about 30% more acidic today than pre-industrial times. Such changes in ocean chemistry can have devastating consequences for the planet's entire ecosystem.

We find such evidence compelling that human activity is causing climate change. Some historians of science have speculated on what caused such a sudden break. One suggestion is that the break in 1769 corresponds to James Watt's patenting of the first practical steam engine, which made steam power cheap and available. An alternative explanation is a change in the banking system in London, England. After 1757, banking became an increasingly effective way of spreading risk. This let the industry grow dramatically even while the national debt grew. The increased industrial activity let London overtake Amsterdam; in 1700, Amsterdam had been larger.

Whether caused by Watt or by bankers, the result is the same: human-made increases in CO_2 concentration have heated the atmosphere of the entire Earth and are making the oceans more acidic.

9.2 A Decarbonized Path Forward

If we conclude that climate change is real and could be mitigated, what can we do? Many experts are using models of the Earth's climate to give estimates of the carbon budget left to limit global temperature rise to 1.5 °C or 2.0 °C above pre-industrial levels. The common answer is that we need to get to zero emissions as quickly as possible. Unfortunately, the world has done the opposite, with record-breaking emissions practically each year. Still, given the remaining carbon budgets, experts can offer potential pathways to decarbonize.

A possible pathway is to minimize the cost of achieving decarbonization using integrated assessment models. As input, these models use estimates of current and future energy generation costs, carbon reduction costs, energy storage costs, energy usage now and in the future, resource constraints, and current and potential future regulations. The models then select the set of technologies that minimize the total cost of achieving energy decarbonization by a given timeframe. Such models have many assumptions—what future technologies are possible, what they may cost, how we consume energy during the day and during the seasons, and what resources are available to generate energy locally and regionally. We believe that it's best to consider results as scenarios, not predictions, of how to achieve decarbonization at the lowest projected costs.

The International Energy Agency recently conducted one such decarbonization scenario. Their analysis is given in Fig. 9.2. The entire world can use a set of energy technologies that can reduce global CO_2 emissions to zero by 2050 and limit the global temperature rise to 1.5 °C (International Energy Agency 2023). This set of energy technologies includes unabated CO_2 emissions from fossil fuels, abated fossil fuels emissions using CCS, direct air capture, nuclear, hydroelectric, wind, and solar, all being within the portfolio of energy options.

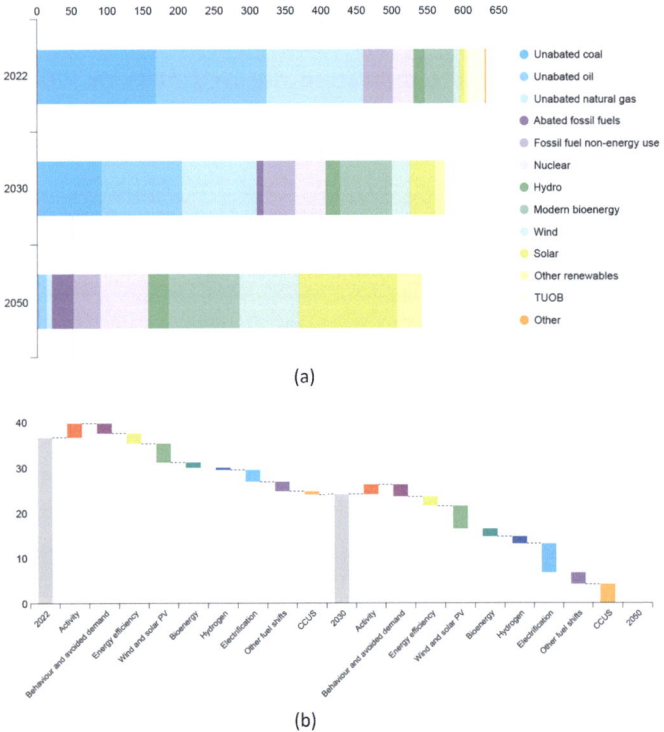

Fig. 9.2 Energy Generation Pathway to 1.5 °C and Net Zero by 2050 (IEA 2023). Integration Assessment Models can provide scenarios that result in **a** lowest cost energy generation to **b** achieve net zero emissions

There are several conclusions implied by Fig. 9.2. First, many wide-ranging technologies are needed to achieve net zero. Just focusing on a few won't get us there—at least not at the lowest cost and in the time frames needed. Different countries, regions, and industries will have different lowest cost pathways for achieving low-carbon energy production.

Second, energy consumption needs to be reduced from 630 EJ in 2022 to 545 EJ in 2050, primarily by conservation. This is a challenging task as the world has historically increased energy consumption, not decreased it. Third, the shift from over 80% carbonized energy in 2022 to nearly 100% decarbonized energy in 2050 must be done at an unprecedented scale and speed.

If the world acts according to this scenario, energy-related CO_2 emissions should drop from 36.9 Gt CO_2 in 2022 to 26.1 Gt CO_2 by 2030 to net zero emissions by 2050, as shown in Fig. 9.2. We want to stress, again, that these models offer decarbonization scenarios at the lowest cost, subject to many assumptions made in the model. Whether such changes proposed by the models are adopted by society is a matter of global cooperation, public policies, and technology development along their projected economics. But one thing is certain. The longer we delay action—which we have by many decades now—the more expensive the pathways become and the more drastic the technologies needed.

Even with a commitment towards building this new infrastructure, our task ahead is immense. Table 9.1 shows the infrastructure needed to supply just 50 EJ of energy using different technologies. For each 50 EJ of low-carbon energy, we would need 175 hydroelectric dams, each producing around 18 GW; or 2.3 million windmills; or 6500 solar parks. Providing the total energy needed will be very difficult.

Table 9.1 Infrastructure needed to make 50 EJ of low-carbon energy by 2050

Option	Size, MWe	Availability (%)	Quantity needed for 50 EJ	Build each year for 25 years
Hydroelectric	18,000	50	175 dams	7
Nuclear	1000	90	1800 plants	70
Solar park	1000	25	6500 parks	250
Wind turbines	2	35	2.3 million turbines	90,000
Solar rooftop	0.004	20	2 billion roofs	80,000,000

A large change in global infrastructure will be needed to enable the energy transition

9.3 Energy for the Individual

So far, we have explored how our entire society will require social restructuring in how energy is obtained and used. We understand that the costs of this transition are significant, but the costs will be higher if we don't act. The longer we wait, the higher the cost, not just in economic terms but also in societal and individual terms. In the previous section, we discussed societal changes. In this section, we suggest individual changes in our private homes and our lifestyle, including our food.

9.3.1 Saving Energy at Home

How we currently use energy in our homes is illustrated by the typical values in Fig. 9.3. Obviously, the greatest single use is heating our rooms and our water. We also spend a significant amount of energy on air conditioning and refrigeration. Other factors are less major, though still significant. The curious wedge in the chart "Adjust to SEDS" represents the uncertainty in all these data and should be spread over all sectors, but since these uncertainties are small, this redistribution will not change the general conclusions much.

Because three-quarters of our domestic home energy use is for heating and cooling, this is obviously the place where we can start conservation. One obvious change is lowering our thermostats. If the outside air is 0 °C (32 °F), and we want the inside air to be 22 °C (72 °F), the amount of heat we lose will be proportional to the temperature difference (22-0). If we reduce the room temperature to 20 °C, we reduce the temperature difference to (20-0), and heat loss will drop by 10%. If wearing sweaters during the winter months makes us uncomfortable, we can consider heating only the rooms of the house where we live. We don't need to heat the patio or the spare bedroom all the

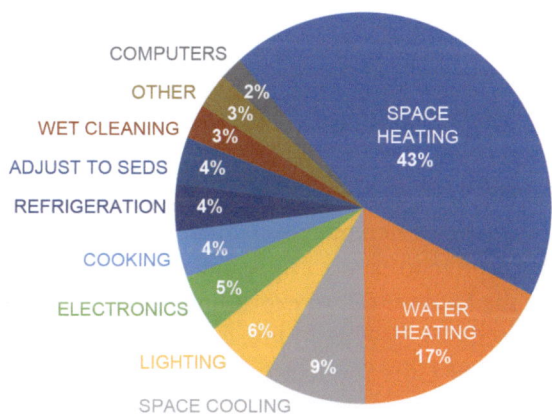

Fig. 9.3 Domestic Energy Use in the United States. About 75% of the energy used in our homes is for heating and cooling (DOE 2024)

time. Other changes, like sealing windows or insulating our water heaters, make similar gains. This isn't hard; it is often incentivized by government and utility rebates and well-detailed in popular publications.

Other energy-saving actions are less obvious. We should avoid electrical resistance heaters because they use work (high-grade electrical energy) to make heat (low-grade thermal energy). Whenever possible, we should use heat pumps because they use work (high-grade electrical energy) to produce about three times more thermal energy than a resistance heater, as was shown in Chap. 4. We should wash clothes in cold water: 90% of the energy used in doing the laundry is for heating the water. Detergents really are that good. Cooking, electronics, refrigeration, and other uses are each 5% or less. Using high-efficiency appliances, high-efficiency lighting, and turning off unnecessary electronics will help.

Reflection: What is the easiest way for you to double your own energy usage? Does this doubling suggest ways to reduce it?

9.3.2 Saving Energy at Dinner

We can also save energy by eating different foods, but this is for somewhat different reasons than we may expect. The energy in our food is less important than the energy used to make the food we eat. In explaining this, we will talk not only about the energy use in kilowatts, but also in terms of the amount of CO_2 produced, because this is more vivid for many of us.

First, we review how much energy and CO_2 are involved. The average American currently uses 10 kW per person and produces an average of 15 tonnes of CO_2 per year, or 40 kg per person per day. This is mostly from industry, transportation, and electricity generation, and not from food itself. If each person has an 8700 kJ (2080 kcal) diet, which is what is normally recommended, food energy per time is only 0.01 kW, a small fraction—about 1/10 of 1%—of average energy usage of 10 kW. From the 8700 kJ of food we eat, we exhale only one kilogram of CO_2 per day, which is less than 3% of the total per person that society emits. The amount of CO_2 directly produced by eating would seem small.

But making food produces much more CO_2 than preparing food. The amount of CO_2 produced to make the food—by growing and harvesting, and preparing it—corresponds to about two tonnes per year per person. Each person is responsible for 0.7 tonnes of CO_2 emitted in making the fertilizer that goes into the food. The fertilizer, which is largely ammonia, is currently made from the combustion of fossil fuels, especially natural gas. In addition to the CO_2 produced as a byproduct of synthesizing the fertilizer, an additional 0.4 tonnes per person comes from growing meat and dairy; 0.3 tonnes per person comes from transporting the food from farm to consumer; and 0.6 tonnes comes from packaging, processing, and waste. Thus, the two tonnes of CO_2 produced to feed each person per year come more from our production methods than from the actual food.

How can we change this situation? First, we could stop the use of fertilizer, which would eliminate the largest source of CO_2 connected with food. But not using fertilizer implies condemning about a billion people to starvation. This is not an acceptable strategy.

A second strategy would be to start making fertilizer from sustainable energy instead of natural gas, from solar or wind-generated electricity. This is saying we should make "green ammonia" from sustainable energy sources rather than "black ammonia" from fossil fuels. This strategy could be implemented, but the costs to make green ammonia using "green hydrogen" from water electrolysis using sustainable energy remain high. A cheaper option currently is to use fossil fuels with CCS to make "blue hydrogen" and then "blue ammonia."

The third strategy for reducing the environmental impact of making food is to watch what we eat. Different foods have dramatic differences in the amount of CO_2 emitted because of their production, shown in Table 9.2. This table shows the amount of CO_2 produced by growing one kilogram of various foods. Clearly, the metaphorical elephant in the dining room is beef, which releases around 60 kg—more than 100 pounds—of CO_2 per kilogram of beef. Lamb and cheese also produce substantial amounts, but bananas and beans are responsible for relatively little CO_2 production. What we eat affects the amount of CO_2 we produce—mostly from production, not consumption.

Individual diets produce different amounts of CO_2. A meat lover's diet produces the most, and a vegan's produces the least. But this is not really the only command to eat differently: we are already urged by nutritionists to eat less

Table 9.2 The carbon footprint of food

Food	CO_2 or other greenhouse gas, kg gas/kg food
Beef	60
Lamb	24
Cheese	21
Pork	7
Poultry	6
Fish (farmed)	5
Beans	2
Bananas	0.7

Meat has a higher CO_2 footprint than plant-based food

red meat for better health, and the differences between our eating a vegetarian diet and a vegan diet are not as big as the differences between eating lots of meat versus lots of vegetables. Moreover, remember that what we eat still produces much less environmental impact than how much we drive cars, or how much we heat and cool our homes. Still, by careful eating, we can reduce this footprint.

Reflection: Design food diets that use less energy.

9.4 How to Test What You've Learned

Before we give the conclusions of this book, we want to test what we hope you have learned. To do so, we summarize the gains promised by electric cars, which have no tailpipe and hence claim less pollution. Are gasoline engines a problem solved by electric motors?

We begin this discussion by reviewing carbon dioxide emissions. A surprisingly large fraction of carbon dioxide emissions comes from daily commuting. For example, assuming a 25 miles per gallon car, commuting 30 km each way requires around 5.6 L of gasoline per day. The gasoline has an energy content of about 34 MJ for each liter. The power used averaged over time equals

$$(5.6 \, \text{L/day})(34 \times 10^6 \, \text{J/L})(1 \, \text{day}/(24 \, \text{h})(3600 \, \text{s})) = 2 \, \text{kW}$$

If you use two gallons of gasoline each day throughout the year, the resulting 2 kW power is 20% of your total use. A bit more driving and traveling will bring this value closer to 30%, consistent with Chap. 1.

We also use energy in airplanes. While the average car in the United States gives an average of 25 miles per gallon, or 10.6 km per liter, a jumbo jet with 300 passengers has an average of 60 miles per gallon per person, or about 25.5 km per liter per person. One trip per year from Minneapolis to London is 4000 miles one way, or about 13,000 km return. A jet can just make this round-trip with 150,000 L of fuel, with an energy of around 36 MJ per liter. If the plane carries 300 people, the power cost per person averaged over the year is

$$150,000 \text{ l}(36,000 \text{ kJ/l})/(3600(24)365)\text{s})/(300 \text{ people}) = 0.8 \text{ kW}$$

Just one European trip results in another kilowatt of power per person, averaged over the year. If we fly enough to qualify for the lowest tier of a frequent flyer program, which characteristically requires 25,000 miles or 40,000 km, we will then have an additional power use of around 2.5 kW. Transportation emissions do add up quickly. Indeed, the transportation sector is the largest CO_2 emitter in the U.S.

But will electric cars dramatically reduce CO_2 emissions and hence mitigate global warming? Electric cars do not have an exhaust pipe belching CO_2. However, if the electricity is made from coal or natural gas, the CO_2 will still be emitted at the power plant. We need to question carefully whether electric cars will save that much.

They do, but the picture is not simple, as shown by the data in Fig. 9.4 (Carbon Brief 2020). Each bar in this figure plots CO_2 emissions from the tailpipe (teal), from the fuel manufacture (orange), from making the car (dark blue), and from battery construction (bright blue). The first two bars on the left-hand side of this figure are average emissions from a car in Europe and from Toyota's Prius hybrid. The emissions from a Prius are about 40% less than a conventional car, a real savings.

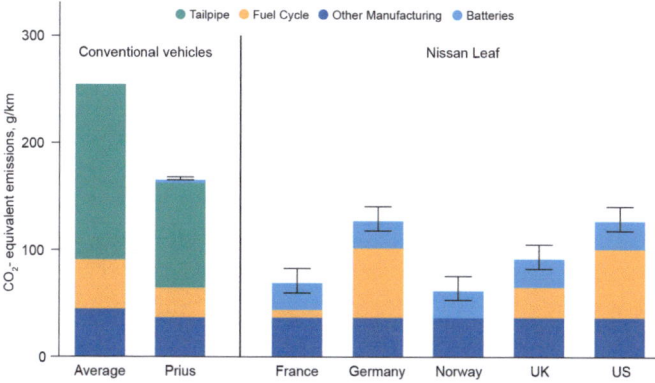

Fig. 9.4 Emissions from gasoline-powered and electric cars. The environmental impact is more complicated than first expected (Carbon Brief 2020)

The details of these two bars are also instructive. The most direct difference is the amount of carbon dioxide emissions from the conventional vehicle vs. the hybrid. Moreover, because the hybrid is more efficient, the emissions produced from making the hybrid's fuel are less because the amount of fuel needed is less. The emissions caused by making the batteries—the bright blue lines—are minor. Hybrids make sense.

The results for a completely electric car, in this case a Nissan Leaf, are shown to the right of the vertical line in the figure. The emissions are less for the all-electric car. However, the picture is not as clear as one might expect. The emissions caused by making cars—the dark blue—are constant among the various countries shown. This makes sense because it is the same car involved. The emissions from making the batteries are constant because these are the same batteries. So far, there are no surprises.

The surprise comes from the difference in emissions. The emissions for France and Norway are small; those for Germany are much higher. What is going on?

The difference is that electricity is made from different amounts of fossil fuels. In Norway, the energy is hydroelectric: all that rain flows into all those fiords to make electricity. France is also in relatively good shape because much of their electricity comes from nuclear power. This is the residue of Charles de Gaulle over 60 years ago, making a concerted effort to make France as independent of fossil fuels as possible. In contrast, the electricity in Germany and in the U.S. depends far more on fossil fuels. A decade ago, the situation for Germany was not as discouraging because the Germans were generating electricity from nuclear power. Their desire to avoid nuclear waste sometimes conflicts with their desire to reduce carbon dioxide emissions. Nonetheless, electric cars make some sense. But what are your own conclusions?

Reflection: At present, electric cars are best charged at night, because that's when fossil-fired steam turbines produce excess power which is cheaper. In the future, electric cars could be recharged with excess solar power available during the day. What does this imply?

9.5 Conclusions

The earlier chapters of this book show that energy consumption is basic to how society lives, both in developed and developing economies. This demand for energy has been sated by burning fossil fuels, which is causing climate change—including a hotter planet, rising sea levels, ocean acidification, and the frequency of extreme weather.

To be sure, we need to conserve energy, but conservation alone cannot solve the climate problem. As we noted in Chap. 1, the need for energy is large. Heat-to-work machines, perfected since the industrial revolution, have ushered in immense global prosperity and improved the human condition. But these machines have an ever-increasing cost. Our society must change urgently and rapidly to reduce this cost.

Just about every year, the world's energy consumption continues to increase, so the challenges are increasingly difficult. We can see this by looking at Fig. 9.5 (Data from Energy Institute 2024). Global energy consumption has been increasing about 2.5% annually for decades. Renewables have not been. Global energy consumption grows, so does our consumption of fossil fuels, and our CO_2 emissions keep increasing. We have run hard, but we are only standing still.

In the U.S., the picture is slightly different. Energy consumption has been flat for the last few decades, but is now expected to grow fast due to the advent of data centers for artificial intelligence. Because of fracking, natural gas consumption has grown while coal consumption has declined. As a result, U.S. CO_2 emissions have declined. Renewables have helped this decline in CO_2 emissions, too. Remarkably, both global and US fossil consumption is over 80% and renewable consumption is under 10%.

The developing world is increasing its energy consumption as well as its CO_2 emissions, considering the historical and other context. Global CO_2 emissions since 1750 have been estimated to be about 1700 Gt, of which the US has emitted nearly 25%, with the combined EU and UK following closely at 22%. China is at nearly 14%, and India is at 3% (OWID 2024). These historical emissions are remnants of countries becoming developed on the back of

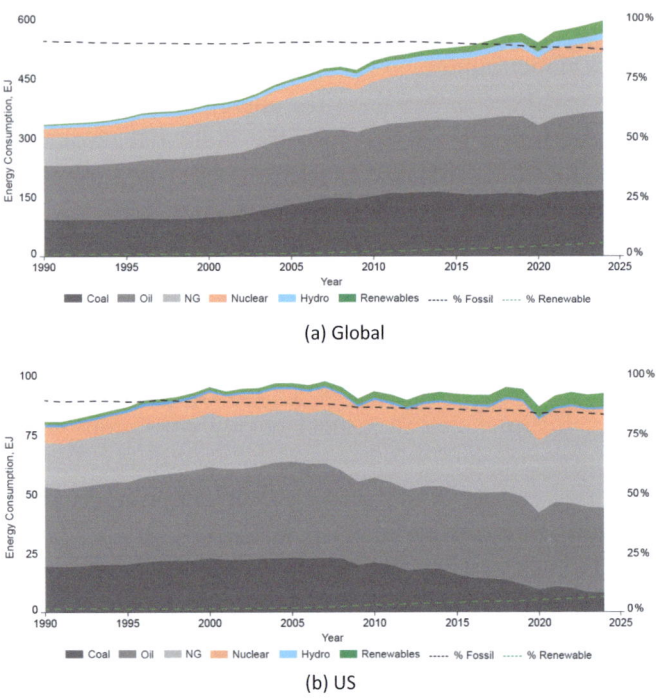

Fig. 9.5 Global and US energy consumption. a The rise in global energy consumption has outpaced the rise in global renewable energy. b The US energy consumption has been largely flat for decades, with increasing amounts of renewables. Overall, both global and US energy is over 80% fossil fuel (Data from Energy Institute 2024)

fossil fuels. Even now, U.S. energy per person is more than five times that of countries like China. As large population countries like China and India develop their economies, their energy consumption per capita will almost certainly rise, too. Not only must the world decarbonize the existing energy system, but we need to ensure that future growth is also decarbonized.

Much of the U.S. and world public is now enthusiastic about developing more sustainable power, especially from the sun and wind. To provide the U.S. with enough energy from the sun, we will need solar modules or wind turbines covering a surface area of around 70,000 km^2. For comparison, Delaware is 5000 km^2, Connecticut is 13,000 km^2, Minnesota is 200,000 km^2, and California is 400,000 km^2. We will need a lot of land. But we will also need dispatchable and baseload power, some of which will be nuclear and some will be fossil with CCS. In this energy transition, we need all options.

We also need power generation to be near the primary consumers, who live along the coasts. This explains why so many are interested in generating electricity from offshore wind. To be sure, electricity from offshore wind is three times more expensive than electricity generated from onshore wind, but we do not then have the problems of moving the electricity from the Great Plains of the Midwest to the urban centers on the coasts.

Reflection: If the world had agreed on environmental action in 2000, would it have made a difference?

This need for urgent need for change is not a surprise, and each year the call for action becomes more urgent, more desperate. In 1896, the Danish physical chemist Svante Arrhenius predicted how CO_2 in the atmosphere impacts the global temperature. Though his predictions were controversial, he was well respected enough to earn a Nobel Prize in 1903. We have been studying climate change for more than 130 years now, when only 10% of the hydrocarbons had been burned as compared with today. Over the years, science has gotten more definitive, human activity requiring energy has increased more than tenfold, and there is high confidence that this activity is causing climate change.

What do you think we should do?

References

Carbon Brief. 2020. *Factcheck: How Electric Vehicles Help to Table Climate Change*. Accessed 1 Dec. 2025. https://www.carbonbrief.org/factcheck-how-electric-vehicles-help-to-tackle-climate-change/.

International Energy Agency. 2023. *Net Zero Roadmap: A Global Pathway to Keep the 1.5 °C Goal in Reach*.

National Oceanic and Atmospheric Administration (NOAA). 2024a. *Climate Change: Atmospheric Carbon Dioxide*. Accessed 1 Dec. 2025. https://www.climate.gov/news-features/understanding-climate/climate-change-atmospheric-carbon-dioxide

National Oceanic and Atmospheric Administration (NOAA). 2024b. *Ocean Acidification*. Accessed 1 Dec. 2025. https://www.noaa.gov/education/resource-collections/ocean-coasts/ocean-acidification.

Our World in Data (OWID). 2024. *Cumulative CO_2 Emissions*. Accessed 1 Dec. 2025. https://ourworldindata.org/co2-and-greenhouse-gas-emissions.

The Energy Institute. 2025. *Statistical Review of World Energy*, 74th ed.

U.S. Department of Energy (DOE). 2024. *Why Energy Efficiency Matters: Energy Saver*. Accessed 1 Dec. 2025. https://www.energy.gov/energysaver/why-energy-efficiency-matters.

MIX
Papier aus verantwortungsvollen Quellen
Paper from responsible sources
FSC® C105338

If you have any concerns about our products,
you can contact us on
ProductSafety@springernature.com

In case Publisher is established outside the EU,
the EU authorized representative is:
**Springer Nature Customer Service Center GmbH
Europaplatz 3, 69115 Heidelberg, Germany**

Printed by Libri Plureos GmbH
in Hamburg, Germany